# Taming the Beast

# Taming the Beast

## Choice & Control in the Electronic Jungle

Jason Ohler

TECHNOS PRESS

BLOOMINGTON, INDIANA

Published by TECHNOS Press of the Agency for Instructional Technology
Box A, Bloomington, IN 47402-0120
© 1999
All rights reserved

"Introduction: 'Uncloaking the Human Option'" © by Paul Levinson,
Ph.D. (1999)

First Edition

ISBN 0-7842-0873-5

Library of Congress Catalog Card Number: 99-64704

Inari, *Book Design and Composition, Cover Design*
Brenda Grannan, *Cover Illustrator*
Kenneth Goodall, *Editor*
Mardell Jefferson Raney, *Senior Editor*
Carole Novak, *Manager*, TECHNOS Press
Michael F. Sullivan, *Executive Director*, AIT, *Publisher*

Printed by Thomson-Shore, Inc., 7300 West Joy Road, Dexter, MI 48130-
9701, USA

*To my wife, Christine, whose patience, insight, friendship, and love continue to mean so much. And to Madeline and Amanda, who make me proud and joyous to be a parent.*

—The Author

*Look and you will find it. What is unsought will go undetected.*

—Sophocles

*there is absolutely no inevitability as long as there is a willingness to contemplate what is happening*

—Marshall McLuhan

# Contents

# Introduction

# "Uncloaking the Human Option"

*by Paul Levinson, Ph.D.*

Pessimism about new media and technology has become the easy coin of academics. Depicting humans as hapless, unknowing, irretrievable victims of our shining inventions instantly distinguishes the critic from the corporate executive—an ever-popular target—and endows the scholar with a real-world visa from the ivory tower. Never mind that such criticism is unnecessary if true—why bother to warn us of a danger beyond our control?—and never mind that it is less than half the picture.

Jason Ohler minds, and makes a point in *Taming the Beast* of presenting the other half: to wit, given that new technologies are intrinsically riddled with unintended consequences, to what extent can we nonetheless move to control them for our general betterment? Oh, Ohler gives due consideration to doomsayers of our technological age from Sven Birkerts to Langdon Winner (frankly, more than I would) but he also consults the more balanced and historically savvy assessments of James Burke, Arthur C. Clarke, Henry Petroski, and most of all Marshall McLuhan in alerting us to the opportunities for human improvement inherent in communications media.

Early on in *Taming the Beast* you'll find a refreshing discussion not only of television's oft-chronicled discontents but also its benefits. "It provides the best of both worlds by allowing experience at a distance," Ohler aptly notes, for it offers us "electronic access to a dominant culture while remaining in a traditional community." He then goes on to suggest and assess strategies to derive the best from this medium. One that I have recently begun to urge is television as an antidote for what ails on the Internet: sometimes it is good to just lay back and relax and let the images flow by you without having to point and click at Web pages.

Ohler goes on throughout this important book to uncover and develop similar strategies for a variety of media, especially those that are increasingly coming into play via the Internet. Does the new technology meet an already existing need, or create one where none existed before? What new problems might it engender even as it remedies existing ones? Electronic books, Ohler observes, "save lots of paper and many trees . . . reduce the burden of carrying heavy printed books," even as they may "increase eyestrain due to poor resolution."

Underlying the very posing of such questions and answers is an essential recognition that humans have options—meaning we, not our technologies, are ultimately in control. This understanding and its application make *Taming the Beast* a crucial manual for the new millennium, a sourcebook not only of good ideas—of which there are many in this book—but even more significantly of an attitude that will help us make the most of the new technological, fundamentally more human, world we are almost daily bringing into greater being.

*White Plains, New York*
*July 1999*

Dr. Levinson is the author of *Digital McLuhan: A Guide to the Information Millennium* (Routledge, 1999).

# First Words

All educational ventures seek to answer a few fundamental questions: What is important to us as a culture? How do we impart it to our children? How do we support it as adults? The questions are simple, but answering them takes a lifetime.

In 1984 the appearance in the Juneau School District of six Apple IIe computers — the very earliest desktop computers to establish a lasting presence in K-12 education — initiated an attempt to address those questions. It was the start of a journey that will never and should never end.

My colleagues and I at the Center for Teacher Education at the University of Alaska in Juneau decided that those six first-generation desktops constituted a trend, and we set about trying to prepare for it. While the future seemed uncertain, it was obvious to us that computers were on the march and were destined to become cheaper, faster, more powerful, more affordable, and more abundant.

Although none of us had any idea just how rapidly Information Age technology would evolve, we did sense that we had crossed an invisible historical divide of immense significance and that there would be no turning back. As we stepped over the line we had the same feelings of excitement and anxiety that accompany a child's chugging climb up the first hill of an unfamiliar roller coaster ride.

That same year a group of us began discussions about creating a program devoted to understanding and applying the new technology toward the business of learning and teaching. Our team consisted most notably of Mary Beth Darrow from Apple Computer, Dr. Joseph Little from computer sciences, and Dr. William Demmert, whose brainchild as Dean of the Center for Teacher Education was to formally introduce educational technology as a discipline at the University of Alaska.

Because we were educators, what concerned us most was the impact of the new media on learning. As we watched high technology find its way into classrooms, we noticed something that would define the Information Age from its earliest days: the infusion of technology produced, in Neil Postman's terms (in his book *The End of Education: Redefining the Value of School*), an ecological rather than an additive impact on our institutions and processes of learning.

That is, adding technology to conventional educational systems did not produce technologically assisted conventional education. It produced something profoundly new. Everything, from classroom dynamics to the roles of teachers, where we found our information, what we studied, and how we paid for it all, was destined to change. And the change was destined to have an immense impact not just in our schools but also in our workplaces and communities.

## Using Technology Creatively, Effectively, and Wisely

Like most groups faced with new enterprises, we at the Center for Teacher Education went in search of a vision and a mission statement to guide our program. We found ourselves in agreement on two basic premises.

First, we agreed that the heart and soul of the program should be empowerment of the average teacher rather than the teacher with a technical bent—the "teachie" rather than the "techie," as we were wont to say. We therefore avoided focusing our conversations on programming and product development, as was typical in those days. Instead we focused on how to help classroom teachers use technological tools to amplify teaching and learning in ways that would empower them and their students to expand, enrich, and ultimately redefine the educational process.

Second, we agreed that information technology was not something we could address within a closed system. We knew that technology was going to turn up everywhere, in schools, libraries, government and business offices, and homes. Whatever our mission statement was going to be, it had to offer a picture much bigger than any one facet of society. It had to embrace the whole of society.

Our discussions about an appropriate vision for the program went on for many months, even years. But eventually they coalesced into this statement:

> The Educational Technology Program supports a number of degrees and activities designed for practicing teachers, educational administrators, and specialists from fields other than formal education who want to develop technical, instructional, and leadership expertise in the field of educational technology, and who want to learn how to use that expertise effectively, creatively, and wisely in pursuit of learning and teaching.

The statement has survived in this form for nearly fifteen years, largely because of three timeless words that are its heart and soul: *effectively, creatively, wisely.*

Using technology effectively in education proved to be a

fairly straightforward concept, though devilish in the details. Teachers needed to know how to operate new technological tools and integrate them into the educational process in ways that produced results.

The emphasis on creativity was our way of acknowledging that computers were fundamentally different tools. They would not just enhance education but also change its very nature by appealing to and facilitating teachers' and students' imaginations in ways that were profoundly new.

But highlighting the importance of wisdom in our use of technology in education was by far our happiest accomplishment. Early on we saw that the new tools were so powerful that using them would require a heightened sense of responsibility as we navigated into the future. We knew that ultimately, beyond the hype and the hope, the glitz and the promise, these tools would enhance our selves as a community — or diminish us. Determining which direction the technology would take us was the domain of wisdom.

It is with the wise use of Information Age technology, not only in education but also in the workplace, the home, and the community, that this book is concerned.

Chapter One explores the tension created by our emotional versus our rational relationship with technology. I examine what bothers us about technology and how we can expand our cultural literacy to include seeing our technology with a new awareness. As a combination of seeing and evaluation, this new awareness will allow us to regain some of the control that eludes us in our rapidly evolving technological world. This new awareness should drive how we redefine our schools and who we want our children to be when they graduate into the world of work and community.

Chapter Two constitutes a technology assessment guide, a

resource similar to the one I have used with my students at the University of Alaska for many years. The reader, assuming a position as agent of the fictitious Science and Technology Administration (STA), is led through a series of questions about technology in order to determine for him- or herself its effects on society. The reader/agent is required to examine technology holistically from three perspectives: functions, goals, and impacts. I have adapted the process from the work of the U.S. Food and Drug Administration (FDA), whose job is to determine the worthiness and safety of new foods and drugs before they are allowed to reach the market.

In Chapter Three I take the reader through an actual assessment of a relatively new technology that could engender its own educational revolution: the electronic book.

The final chapter, Last Words, addresses the important matter of how all of us can imbue our approach to Information Age technology with balance, with informed viewpoints, and, let us hope, with the wisdom necessary to take back the future.

## Acknowledgments

I would like to thank the many mentors I have worked with throughout my life, some actual, some virtual, and many of whom I have never met in person and know only as words on a page or screen.

I would like to acknowledge Dr. Bill Richards from Simon Fraser University, whose dedication to helping me navigate the mysteries of graduate schools is forever appreciated.

I owe much to Mary Beth Darrow, who started me on the educational technology road and who, more than anyone else, helped me see why understanding technology was essential to being a good educator. And I owe much to Dan Malick, who helped me understand just how powerful a tool a computer can be when applied intelligently to the world of human endeavor.

I would like to thank the University of Alaska Southeast, in particular Dean John Pugh, Vice-Chancellor for Academic Affairs Dr. Roberta Stell, and Chancellor Dr. Marshall Lind, for helping to provide the time and resources necessary to make this book possible.

I would like to thank Ken Goodall for his patience and wisdom as my editor, Mardell Raney of TECHNOS Press for understanding the importance of this work and supporting it throughout its development, Mary Bowen of the University of Alaska Southeast for her many suggestions and hours of editorial help in preparing the manuscript, and Margaret Grogan, also of the University of Alaska Southeast, for help in transcribing too many hours of tapes and for proofreading and general support.

And I would like to thank my parents, both of whom were teachers, and both of whom instilled in me at an early age the value of curiosity coupled with compassion and purpose.

# One

## The Mirror in the Machine

We like to think of technology as something external to ourselves. Doing so helps us feel we are in control. And we like to think of it as a rational construct. That helps us feel we can apprehend and contain it through science, mathematics, engineering, and the other languages of the intellect.

But in reality technology exists as much within us as it does outside us. Like two people living in close quarters, we begin to assume each other's personality.

And technology is as much an emotional experience as a rational one. It externalizes not only our logic and thinking but also our dreams, feelings, prejudices, and fears.

That makes our technology a much more accurate mirror than we ever suspected. If we look closely at our machines we can see ourselves—all of ourselves, complete with the anxiety, creativity, problems, and opportunities of being human.

The tension created by emotional versus rational experiences of technology dominate much of this chapter.

We examine the problems with technology that bother us, scare us, and challenge our humanity.

We look at the philosophical underpinning of technology—

the propensity to subvert tradition—and how it challenges so profoundly the mythology of schooling that typically has provided so much of our cultural identity.

And, I hope, we will see that if we are willing to expand our notion of being literate to include reexamining the technology we take for granted, we may ultimately regain some of the control many of us feel we have given away to our machines and impart a new sense of social responsibility to our schools, our children, and ultimately our communities.

## What Technology Isn't—and What We Are

Technology is never just about itself.

Technology is always so much more than we think it is. It is an arena, a lens, a totem etched with legends that each storyteller tells differently, a high-resolution mirror reflecting our blurry, vibrating cultural personality, and so much more. But it is never just itself.

As with all things not themselves, technology is a metaphor. Robert Frost, in an essay entitled "The Constant Symbol," wrote in his no-frills way that metaphor is "saying one thing and meaning another." We say *this* in terms of *that*.

Technology is how we try to speak to ourselves and each other in the tongues of aspiration, through images of our becoming: saying mobility in terms of car, community in terms of television, individualization in terms of cubist office environments.

*Technology is a measurement of the distance between two cultural reference points.*

Technology is the conversation we cannot have directly with ourselves because we haven't mastered the ways of inner dialogue or the methods of the unmediated experience. It is how we imagine out loud.

## Imagination Outpaces It

Our imagination always outpaces our technology.

The distance between the two creates the gap that the creative spark jumps in order to ignite our forward momentum. Like a sprinter eyeing the next hurdle, we are enticed by the gap, forever smitten with our tinkering as art, as technique, as knowledge, without much thought about the race we run.

If it is true, as they say, that in the digital domain we can create whatever we dream, and if it is also true, as they ought to say, that in this age of ever more powerful machines we need desperately to understand who it is we want to become, then we should be looking at our dreams rather than our machines.

*Machines don't evolve — dreams do.*

Rather than thinking in terms of the evolution of the machine, we should see ourselves reflected in the machine and ask, Who do we dream of being?

But I am getting ahead of myself. The real question is this: What is important? The rest is scripting.

## What's Important

My community entrusts me with the most important role it can assign to one of its members. I instruct its citizens in the ways of the world vision, belief system, and mythology it continually and often unconsciously reaffirms in order to perpetuate its survival in the manner to which it has become accustomed. That is, I am a teacher.

We do not think about our education systems in such terms because we focus on their quantifiable mandate: test scores, success stories, the balance of happy versus angry taxpayers at school board meetings, the ratio of failed people slipping through the cracks of the system and draining community resources to those providing the resources upon which the community draws. Quantifiable results seem somehow more neutral, more culturally independent, than belief systems because we do not hear a stentorian voice of cultural authority anywhere amid the facts and figures.

Yet the voice is there, embedded in the measurements we use. We just need to look deeper and listen more closely. The subtext of quantifiable results speaks softly; and yet, like the subtext of conversation, it carries most of the meaning.

## Who We Are

There is no clearer projection of what a community considers important than the local school.

While the larger crucible of state and federal law forms the shell and the organizational mandate, the school remains one of the few technological formations with handles that the average community member can grasp through personal observation, participation, legislation, and ultimately voting for or against school taxes, school boards, bond issues, codes of behavior, and other instruments of shaping the current mythology.

School is an elaborately constructed filter through which our young people pass into community life and from which they inherit the measurements of their culture.

There is no clearer sign of what a school is than the measurements it has adopted to calibrate the success-failure continuum. Where a measurement, like a medium, goes unchallenged and assumes the rhythm of self-perpetuation, the measurement becomes the message, revealing the true, though unconscious, aspirations of the measurers.

## What We Believe

We are not different from our schools. All of us, from our clothes to our unconscious, from our measurements to our methods, are walking, talking belief systems whose lives tell the story of who we really are.

If you ever want to know who you really are, rerun the events of a typical day and ask yourself what belief system must have led to your behavior. You will find enough consistency to infer a coherent mythology uniquely your own yet expressive of the larger culture to which you belong—a one-off creation within the confines of a mass-produced world.

Beliefs are by nature beyond explanation. Each belief system, from atheism to animal worship to Roman Catholicism, pushes adherents to a precipice from which they are required to leap into strange and often intellectually untenable territory. When you don't know what you don't know but nonetheless are required to act, you must act based on what you believe. That is how schools behave.

And that is how we behave toward technology. We leap not necessarily because we want to but because we must, because the mythology of modernity makes us afraid of being left behind, of depriving our children of the tools of their time and the pathways to their imagination.

## LiteraSee

So what does a school believe when it comes to technology? What leap does it make because it must?

While the endless shifting of a democratic culture rocks the pedagogical and philosophical pendulum between extremes, we postmodern humans, regardless of belief system, converge on at least one principle of personal responsibility: our children should be able to feed themselves by themselves.

Toward this end our schools rally to the cry of "work readiness." We believe in the need to arm our children with

*It's hard to be yourself; then again, it's harder not to.*

*The future is not ahead of us pulling; it's behind us pushing.*

the skills of upward mobility so they can move forward and outward, away from the immediate communities that nurtured them.

In that belief we differ greatly from agrarian humans, for whom there was no self apart from community, but less so from industrial and modern humans, though the means of attaining the objective have changed.

*All cultures require their young to serve apprenticeship in the ways of the machine. Schools for tools, as it were.*

There is of course a tremendous amount of power and efficacy in disseminating technological skills as widely and deeply as possible in order to make the principle of self-reliance a reality. After all, schooling as we know it was created to raise people in the ways of technology in order to conceive, design, build, maintain, repair, and operate the machines of labor.

Today—whenever that is—is no different.

## Literal Beliefs

Vision and belief within educational systems become concrete when they assume the form of literacies.

The most important task we have entrusted to our teachers is the transmission of literacies—skills, paths, belief systems. Literacies are the hands of the ubiquitous, invisible, knowing masseuse, molding the collective internal landscape and polishing the lens through which we project the world we expect for ourselves and for each other, particularly our children.

*Webster: Literacy is the ability to read and write.*

Teacher, citizen, student with citizen-in-training status— all are sculpted by the literacies which the dominant culture has identified as critical to everyone's success. They are everyone's insurance policy.

We know them as the 3 Rs. But this tag is purely habit, a routine of mind and perception mythologized by a mechanical culture, thoroughly and iconoclastically challenged by the new media.

In the absence of mechanics as the dominant cultural metaphor—replaced for now by the associative meanderings of life on the World Wide Web—the 3 Rs are free to evolve. If the notion of "reading and writing" is deconstructed to its essential elements of "coding and decoding the symbols of the dominant culture of the day," then Webster's definition works well for print, multimedia, virtual haptic environments, and whatever else we might invent.

With the advent of the Web, foundational literacy has expanded from the 3 Rs to the 4 Rs. In an age of multimedia, in which we are expected to communicate in pictures and sounds as well as words and sentences, in graphic splashes as well as linear, interchangeable symbols marching from left to right, art looms as the obvious next great literacy, the first "R" to be added to the cultural repertoire in centuries. (See my article, "The 4 Rs: Reading, 'Riting, 'Rithmetic, and aRt Across the Curriculum," in *TECHNOS Quarterly,* Vol. 5 No. 2, pp. 26 & 27.)

If there is one hole in our approach to literacy that bears singling out, it is the fact that art has already arrived as the 4th R and is waiting for us to recognize it. In an age in which the term paper is yielding to the Web page, it is just as important to be literate in the use of pictures and sounds as it is in the use of text. Yet where in a standard curriculum do students receive the art and design training needed to achieve this literacy?

In *The Soft Edge,* Levinson makes the case that monotheism only became possible with the emergence of a phonetic alphabet. Its predecessor, hieroglyphics, used pictures, presenting the problem of creating a picture which described a universal essence—a bit like trying to create a container for a universal solvent. Writing about universal concepts forced us to abandon pictures in favor of a language based on speech to enhance our ability to deal in concepts. We now return full circle. In an age in which billions of pages of material flood the Web every day, using a picture to say a thousand words—or a video to say a million—offers a promising respite. In a world gone international because of the ubiquity of the Net, using something less ethnocentric than text—like pictures—begins to make sense of a global sort. We live in a world that can only be partially described with words, and the artist must come to our rescue. (For more information about art as the 4th R, visit the FourthR Web site at http://www2.jun.alaska.edu/edtech/fourthr.)

## Background/Foreground

But the 4 Rs belong to the domain of foreground literacies. It is the background literacies that form the bedrock of culture.

*How to see the forest from the trees? Climb a tall tree and look around.*

The work-readiness mindset often forgets that education should prepare us for more than just work. It should prepare us for life. It should prepare us for community, however virtual or distributed. Through school we become not just workers but also citizens and neighbors.

Communities are much more than derivatives of the machine world guided by efficiency and profit. They are in fact much more than the sum total of our labors. They are multifaceted activity centers, a convergence of play, life, death, education, worship, socializing, and self-discovery—as well as work.

*We teach what we live.*

If we school for work only, then we *become* the machine. If we school for life, we have a better chance of being ourselves with a life outside the machine, of perhaps even imbuing the machine with some of what we dream to be.

While work may demand hewing to the current edge of technological fashion, what we do in our private lives can be anything we wish. Unfortunately, life preparation, if broached in the classroom, is often seen as meddling, challenging the mythologies of taxpaying families and members of subcultures with specific religious and social agendas. Alas, back to work readiness.

*Attitude, not aptitude.*

Background literacy is becoming so important that it is moving to the foreground, where we can focus on it and develop it deliberately and scientifically. The oft-cited Secretary's Commission on Achieving Necessary Skills (SCANS) report, an effort by the U.S. Department of Labor to identify workplace competencies for future workers, lists as many important background behaviors (responsibility, self-esteem, sociability, self-management, integrity, teamwork, problem solving, creativity, visualization) as foreground skills (systems, information, and resource management; technological competency; the 3 Rs).

In an age in which skill sets shift rapidly, who you are is more important than what you know. It is a greater indication of your ability to submit to the relentless churning of the workplace that now defines the very nature of work.

## *Heralds of Community*

*In the transfer of background as well as foreground literacies, teachers are community messengers.*

*Everyone is sculpted by the behavioral literacies that teachers model as a result of being who they are. And who are they?*

*Teachers are:*

- ◪ *Purveyors and examples of literacy.*
- ◪ *Citations of self in the narrative of their culture.*
- ◪ *Citations of culture embedded in themselves.*
- ◪ *The warmware interface between self and others in the cultural amalgam.*
- ◪ *The medium bridging now and future generations.*
- ◪ *Officially sanctioned agents of the dominant mythology.*

## Four Stages of Literacy

But what exactly is literacy? The best way to answer this question is to consider the four stages in which literacy develops.

Stage 1 literacy is seeing. *Seeing* is a relative term, more a metaphor for the many ways to forage for information, to hunt for and gather input. A bat sees with radar, Beethoven with inner ears, an empath with its heart.

As Marshall McLuhan reminds us, preliterate humans saw with their ears.

Stage 2 literacy is absorbing. Absorption is acquisition of the hidden curriculum. It enables students to read their culture like a video program, passing through it like the hypnotic interface of a complex, compelling computer game—as if it were not there. Conscious negotiation of the tricks, traps, and syntax of life disappears, moving to the background. All that is left for the conscious mind to focus on is the final score.

Max Frisch describes technology as a way of organizing the universe so that we don't have to experience it.

Stage 3 literacy is forgetting. To be unaware of your literacy is to be fully acculturated. When you forget that you are seeing and absorbing, you are living fully in the present.

Stage 4 literacy is a value-added stage 1, a return to seeing with a new depth, the postmodern human with a split consciousness, both seeing and evaluating on the basis of new principles of freedom, democracy, and the right to reconstruct tradition.

So we have four stages of literacy:

Stage 1. Seeing
Stage 2. Absorbing
Stage 3. Forgetting
Stage 4. Seeing *plus* evaluating

This book is about how postmodern humans—you and I—can cultivate stage 4 literacy, learning to see, to evaluate, and thereby to control technology.

## Tradition vs. Sedition

Stage 4 literacy — seeing plus evaluating — did not exist among the general run of human beings until recently, historically speaking.

*Nostalgia is the irrational longing for limitations.*

Those who have passed into stage 4 literacy are split into two factions. One faction seeks to improve upon the old ways. Members of the other faction rue the day they passed through the doorway of scientific knowing, as though an *a priori* appreciation of our future could have spared us so much of our current misery and kept our Garden of Eden intact. Their keening presupposes that because today is so confusing and because we are allowed to imagine living in ways that are not, the past holds the key to future happiness — if only we could forget what we know.

But ignorance *a posteriori* is earned, not inherited. We must pass through a state of awareness of the expansive freedoms of the democratic age in order to know that we can choose a life that has fewer choices. Bliss in ignorance is not imposed on us; we must deliberately choose our ignorance for ourselves.

*Complete freedom includes the liberty to choose limitations.*

As we learned from the woman's movement of the past thirty years, a woman who somnambulates into a life as wife and mother because it is expected of her is not free. But a woman who is a wife and mother because she chooses to be *is* free. The result, looking from outside, appears to be the same. But the interior view is radically different.

### What's Natural?

One of the most important additions to our general cultural repertoire during the past millennium is our greater willingness to question accepted, collective tradition. Our propensity to do so, which began as a heretical itch, has flourished into an obsession, even an expectation, among us postmodern humans who pride ourselves on our freedom and our curiosity.

The most poignant struggle for the soul of our culture today is between those who want to use curiosity to nourish the culture and those who feel that curiosity threatens it. This struggle involves a dialectic between what we often refer to as liberalism and conservatism: a natural versus an unnatural curiosity to question the world around us.

In large part this newfound curiosity conceptually identifies the master skill set that characterizes postmodern

## Tradition vs. Erudition

*I grew up in the suburbs in a lower-middle-class Caucasian family, raised by parents who made me aware of racial inequality at an early age.*

*My father fought to racially integrate the private school at which he taught. He was ultimately successful, despite an uphill battle that was worthy of the Sisyphus Hall of Fame. The rock he rolled was change. The hill was tradition. He then fought and ultimately succeeded just before his death in helping to make the same school available to women. The same rock, the same hill, the same rolling.*

*Rock and roll began with Sisyphus.*

*As a result of my upbringing I came to regard longstanding social structures with suspicion. I considered the phrase "a wonderful tradition" to be an oxymoron until proven otherwise. The process my father followed was from tradition through erudition to petition for sedition.*

*No culture is immune to this process. It was not until I moved to Alaska that I became truly aware of the fact that America had a Native population outside the standard Hollywood caricature. At a friend's house in Juneau I studied a map of the United States that outlined indigenous lands, and with disorienting amazement and visceral embarrassment I realized for the first time that the town in western New York where I had spent the first eighteen years of my life was surrounded by a number of Native American reservations and traditional lands.*

*I didn't know. No one had told me. American history class had been a tribute to European expansionism in which Native culture was a troublesome footnote. In history class there had been no mention of anything local. History had always been very far away.*

To most of us,
technology feels
more natural than
nature.

civilization. It encompasses our ability to explode nature into fragments in order to understand it as a pile of minutiae, then reassemble it in new ways we call technology.

In many ways this master skill set underscores everything we teach in school and everything we expect from our children. Our devotion to classroom cultivation of what we call critical thinking skills, which form the attitudinal bedrock of lifelong learning and self-actualization, assumes that nothing is beyond question — except, of course, the dominant cultural mythology.

How different that makes us from the august modern

*The Alaskan Native population thrived on its past. But by the time I arrived in Alaska, Native culture was well into the endgame of trying to salvage what tradition it could from the great juggernaut of progress driven by Christianity, technology, a scientific view of nature, and a cash economy.*

*The assault on Native ways of life had come so quickly and was based on perceptions of life so different that there was no time for the Natives to mount a defense. With irony in full bloom, democratic thought, amplified by mass media — the very antithesis of localized traditional values — produced a collective embarrassment in the dominant culture strong enough to at least minimize state-sanctioned cultracide.*

*The same week that I sat mesmerized by the map of Native lands I gave a ride to an Alaska Native hitchhiker who was interested in combining Western medicine and Native ways of healing. She confided to me that she had had an argument with one of her elders about the traditional role of women in the Native community. She had been studying, thinking, and comparing her way of life against the backdrop of the open slate of possibility. Through erudition, an assessment of tradition became petition — leading to sedition.*

What's tradition to one is sedition to another.

*But like all seditions it would become a tradition over time. The change in the young woman's Native way was not her suggestion that women be allowed to assume new roles but her broaching the subject at all — that she could address tradition as a communication medium.*

*Through efforts like hers, Sisyphus's hill would level out, the rolling would become less arduous, the rock a little less inertial. And then, slowly, the flat ground she helped to create would tilt and become the next hill, as stage 3 literacy — forgetting — once again took control of a people's unconscious.*

humans of yesterday, clinging like T.S. Eliot's J. Alfred Prufrock to the crumbling shoals of tradition. And how disruptive—this desire to have our cake and eat it too, to live a life of scientific rationality while maintaining an unquestioned faith in a mythology of the past that is anything but scientific and rational. This desire is the essence of the postmodern condition, a kind of cultured split personality in which we promote change and stasis at the same time.

We not only live our lives as Marshall McLuhan said, driving forward while looking in the rearview mirror. We also have our foot on the accelerator and brake pedals at the same time. Life may be emotional chaos without tradition, but with it we suffer creative asphyxia.

As a result of my upbringing I came to regard longstanding social structures with suspicion. I considered the phrase "a wonderful tradition" to be an oxymoron until proven otherwise. The process my father followed was from tradition through erudition to petition for sedition.

Once upon a time our process of becoming meant discovering who we innately were in relationship to an established mythology. But we gave all that up when we adopted literacy, trading the immediacy and limitations of reality for the limitlessness that is inherent in the creation of symbols.

Now we imagine new personae and develop ways of getting there. That puts us far outside the program of tradition.

## Traditionally Speaking

Traditionally we turn to tradition to guide us; otherwise it would not be tradition. The tradition of not questioning tradition is what gives communities their stabilizing core of values. It can also be the tautology upon which culture and intolerance feast.

*Tradition—you can't live with it and can't live without it.*

*Tradition is both medium and message simultaneously.*

A teacher of mine once used this narrative to explain how tradition perpetuates culture: A boy says to his father, "Dad, I have a problem," and Dad says, "Son, I know what you mean, because when I was your age I went to my father and said, 'Dad, I have a problem,' and Dad said, 'Son, I know what you mean, because when I was your age I went to my father and said, "Dad, I have a problem"'" . . . and so on.

With our immense curiosity, we have broken this algorithm and replaced it with another, a much more open-ended subroutine that continually asks, "I wonder if . . ." Much of what tears at societies is not a conflict of ideas but the battle between tradition and the right to have new ideas at all, between the process of questioning tradition and generating new ideas at odds with the old ones.

*We mistake hind-sight for insight.*

Yearning for days of less stress and less change through less technology does not address the true source of emotional conflict. One would need to yearn for the days in which our natural curiosity was kept down on the farm. But doing so would deprive us of all the nifty machines that we have—those things that give us the leisure to rethink the efficacy of filling our lives with the machines that supposedly have freed us.

## What Isn't Technology?

The topic of technology is inherently confusing because it points in so many directions, like a map that goes everywhere at once. It would be a good idea to get some definitions under our hats.

Earlier I described technology as a metaphor. Ultimately that is where the journey of this book takes us. But for now let's consider the more common uses of the word *technology*. Like most great concepts, such as government and the economy, technology is so vast that it works at three levels simultaneously: structure, process, and stuff.

## The Three Levels

In *Autonomous Technology* Langdon Winner defined the structure level of technology as "all varieties of technical social arrangements."

When we say that "technology will save us" or "technology is out of hand" we speak of the permanent structure of technology in the abstract. As pervasive as air, it defines not just what we do but also what we are capable of.

It is, in a most complete sense, an ecosystem—or, to use a more precise term, a tecosystem (TEE-ko-system). It circumscribes our behaviors on every level, from the physical to the philosophical. Living in a tecosystem is no different from living with nature, a force which overwhelms its human visitors, causing them to either bend to its will or become unproductive and useless.

On its second, or process, level, technology covers how we use our things, something that Winner calls technique. He defines *technique* as "the whole body of technical activities—skills, methods, procedures, routines—that people engage in to accomplish tasks."

It is at this level of schooling that the drama of the first three stages of literacy—seeing, absorbing, forgetting—is played out. It includes adopting and adapting the machines of our age within the confines of technology-as-system. Through a willing suspension of suspicion we forget our fears and come fully into the fold of the machine in order to become adept at using it.

Winner refers to the third, or stuff, level of technology as the apparatus. On this level, less abstract than the first two, we refer to technology as things ("Doesn't our computer lab have some fine technology in it?" or "In order for employees in my business to be successful they need to be able to use the technology I have").

It is the first level, technology as structure or system, that perplexes most people in stage 4 literacy, the ones who not only have seen, absorbed, and forgotten but also hope to evaluate technology. It is the big-picture consideration of technology that allows us to see it in all its horror and glory, in terms of the promising but uncertain future it guarantees us.

Since you're reading this book, you're probably one of the stage 4 literates—or at least a stage 4 wannabe.

*Technology—to quote Steve Martin, that's stuff like fur-lined sinks and electric dog polishers.*

## Rational or Emotional

It is from the big-picture perspective that we can play armchair philosopher and see the problem with technology. But first we

---

### *Kids & Adults*

*While we're in a defining mode it might be helpful to clarify a couple of other terms.*

*Kids are those postmodern humans who grow up fearless of the technologies that surround them because there is no distinction between their world and the technological world. This is not to say they are open to change; they* are *change and thus are not aware of it. They live in a rapid recycling of literacy stages 1 through 3, seeing, absorbing, and forgetting in one fell swoop. Stage 4, seeing plus evaluating, is simply no fun at all.*

*Ashen-faced adults are those postmodern humans who start out as kids, as defined in the preceding paragraph, and eventually morph into overwhelmed, anxious elders—aware of change, leery of transformation, nostalgic for the past. They see their children's future in terms of their own past, unable to relate the unknowns of their own childhood to the unknowns their children will someday come to know as adults.*

*I still think like a young person, just not as fast.*

---

need to distinguish between technology as rational construct and technology as emotional experience.

A professor of mine once observed that we are drops of reason in an ocean of emotions. This fundamental truth of life creates a strained dialectic in our own thought and speech.

One side of our mouth seeks to express our overwhelming propensity to experience life as a festival of feelings. The other side seeks just as forcibly to express our antithetical tendency to view reality as rooted in the rectifiable chaos of a natural, inherently rational world. Life is what results when the two collide.

But the prevailing sentiment in our postmodern scientific world comes down hard on the side of rationality. Given enough time and resources, it is said, we brain-heavy humans can logically dissect any part of nature and refashion it into something completely new. Or at least we can explain it in enough detail to qualify as an expression of our intellectual dominion.

So how do we best describe what technology is from a rational perspective? Rationally speaking, technology is anything people create. Everything else is nature.

While the rational side of us describes technology, the emotional side reacts to it. And as with all emotional reactions, a subject of great agitation moves into the foreground where we can fixate on it.

Emotionally speaking, technology is anything people create that people notice. Few people experience a hard-disk crash rationally. Most experience it with anger, frustration, disappointment, and feelings of abandonment and betrayal.

Much of our unconscious interaction with technology is positive. Much of our conscious interaction with technology is less so. And much of what we don't know about technology can hurt us or help us, but not in ways we are trained to understand.

## Things That Go Bump

*People notice technology if it is new, has been replaced, has returned as kitsch decor, breaks, or hurts.*

*New technology often takes us so much by surprise that we feel we have been ambushed by the future. To use Michael L. Dertouzos's expression in* What Will Be: How the New World of Information Will Change Our Lives, *"the ancient human being" in all of us is automatically frightened by and suspicious of new technology. We look askance at it until we become convinced it is not a threat or resigned to the fact that there is little we can do to inhibit its presence.*

*Replaced technology includes vinyl records, rotary phones, milk bottles with thick cardboard plugs, black-and-white television, clothespins. Such items become metaphors for a simpler life in which we had fewer expectations about and fewer opportunities for disappointment with our technology. But we don't notice replaced technologies for very long. We are too busy forgetting the technologies that replaced them.*

*Items that are obsolesced or replaced often return as decoration, set in high relief against the postmodern environment. They levitate to the foreground to a high degree, reducing most things around them to wallpaper. A vintage car on a Los Angeles freeway, an old Coca-Cola poster at a computer workstation, a creaky wooden roll-top desk packed with the artifacts of a neoteric office, a canvas canteen hung outside a Gortex tent: all very cool.*

*When a technology breaks, it moves from background to foreground very quickly. Often we are left to feel betrayed by something that lured us into its embrace, cultivated our dependency, then abandoned us just to show us who had the upper hand. Herein lies the great paradox of the technology that we supposedly enslave: we are much more dependent on it that it is on us. We are compelled to use it because without us it seems useless. But when it breaks we discover who really owns whom.*

*Technology that overwhelms, manipulates, and overpowers us, but does so in ways that we approve of either overtly or tacitly through our acquiescence, does not command our attention. Only technology that harms or threatens us—a nuclear power plant, a poisonous waste incinerator, a glut of spam—shows up on our already crowded radar screen. It moves to the foreground quickly, then recedes as soon as the pain resolves.*

**The conscious experience of technology is often negative, so we try to forget it is here.**

*But today's pain is tomorrow's stage 3 literacy. We see, we absorb, we forget.*

# The Problem with Technology

Each one of us is two people: philosopher and philosophee.

As philosophers we stand on the ledge of objectivity, looking out over time, space, and humanity, discerning the long-range pitfalls of technology. It is from this perspective that we render judgments such as "genetic engineering is wrong because it tinkers with basic relationships among humans, nature, and spiritual essence."

*Philosopher vs. philosophee: a matter of me vs. me.*

As philosophees we are individuals in the welter of humanity who appear as just a blur to the philosopher perched on the ledge. As faces in the crowd we forget our theoretical quarrels with genetic engineering when it offers us or someone we love a better life. We can't get enough of it fast enough. We go from global to local in a heartbeat.

## Split Perspectives

Being two people in one, each of us faces in two directions when we situate ourselves in the tecosystem:

*F. Scott Fitzgerald once remarked that the test of a first-rate intelligence is the ability to hold two opposed ideas in the mind at the same time and still retain the ability to function.*

- As philosophers we complain that personal computers make life more stressful, while as philosophees we praise the wonderful labor-saving capabilities of word processing.

- As philosophers we abhor the invasion of privacy that information agencies and electronic eavesdropping represent, while as philosophees we laud their use to track and arrest the bad guys.

- As philosophers we claim that television is a corporately sponsored festival of misinformation, while as philosophees we are fond of talking about the informative or entertaining TV show we watched last night.

- As philosophers we complain that timber companies are

responsible for stripping Mother Nature of her life-blood, while as philosophees we cherish the beautiful mahogany wood that adorns our furniture, art objects, and musical instruments.

Our technology splits us right down the middle. It creates two people out of each of us, then pits one against the other to such an extent that any objective observer might conclude there was a flaw in the design of human nature.

## Rational vs. Emotional

This split occurs in the same two perspectives that distinguish technology as a construct: rational and emotional. The two tables on pages 30 and 31 show how this split occurs in both perspectives in relation to the technology of television—more specifically, the activity of watching television.

What can we conclude from our split-perspective analyses?

Each technology cuts both ways, connecting us to experiences that make us greater and disconnecting us in ways that reduce who we are, liberating us on one hand while enslaving us on the other. Which side we, as inheritors of Werner Heisenberg's uncertainty principle, actually see depends on our perspective.

*Technology connects and disconnects; balancing the two requires stage 4 literacy.*

It is this inherently dualistic nature of technology that creates psychic tension as each of us tries to reconcile the philosopher and the philosophee within us. We are left mostly on our own to attempt to find a balance. The stage 4 literacy skills that would help us make our evaluations remain largely unaddressed in our schools.

When the rational battles the emotional, pleasure usually wins out. In fact, few people are aware of the pros and cons of television considered from a rational perspective. People mostly act emotionally—for pleasure, for fun, for entertainment, for

## TV's *Rational* Value: A Split Perspective

| *Philosopher* | *Philosophee* |
|---|---|
| ◪ TV robs people of their ability to imagine. | ◪ But it introduces us to new ideas that stimulate our imagination. |
| ◪ TV gives corporations control of the population by projecting a single image of reality into the minds of millions of people at the same time. | ◪ But that single image can act as a cultural reference point that facilitates dialogue and insight about universal global issues, bridging diversity in the name of common cause and interest. |
| ◪ TV accelerates the pace of change, particularly for indigenous cultures, to a rate faster than can be responsibly assimilated. | ◪ But it provides a meta-perspective, documenting our changing times so that we might better direct the process by being more aware of it. |
| ◪ TV obliterates cultures by blanketing them with misleading and irrelevant images of who they ought to be. | ◪ But it provides the best of both worlds by allowing experience at a distance: we have electronic access to a dominant culture while remaining in a traditional community. |
| ◪ TV limits interpretations of reality to those advertisers who are willing to pay for it. | ◪ But advertisers are willing to pay for many enlightening programs, from *The Wizard of Oz* to historical perspectives, that help us understood who we are. |

transport from the immediacy of their existence. Their goal is not stage 4 but stage 3 literacy — forgetting.

## What Bothers Us about Technology

Your printer breaks down, moving from the background of your life to the foreground. A stage 4 literacy crisis thrusts itself upon you.

If you look closely you will see, within your stage 4 crisis,

## TV's *Emotional* Value: A Split Perspective

| *Philosopher* | *Philosophee* |
|---|---|
| ▪ TV robs people of their ability to imagine. | ▪ But "Seinfield" reruns are fun. |
| ▪ TV gives corporations control of the population by projecting a single image of reality into the minds of millions of people at the same time. | ▪ But "Seinfield" reruns are fun. |
| ▪ TV accelerates the pace of change, particularly for indigenous cultures, to a rate faster than can be responsibly assimilated. | ▪ But "Seinfield" reruns are fun. |
| ▪ TV obliterates cultures by blanketing them with misleading and irrelevant images of who they ought to be. | ▪ But "Seinfield" reruns are fun. |
| ▪ TV limits interpretations of reality to those advertisers who are willing to pay for it. | ▪ But "Seinfield" reruns are fun. |

a heirarchy of awareness. At the bottom of the awareness hierarchy you simply see the details of the technological dysfunction in order to conquer them and return to stage 3 literacy as quickly as possible — that is, get back to work. At the top of the hierarchy you encounter philosophical considerations of the problems that shape the human condition, of which your malfunctioning printer is just a metaphor.

*"The world in a grain of sand"—did William Blake see semiconductors that far in advance?*

If you find yourself focusing on your printer longer than it takes to fix it, then you are probably in a big-picture mode. If so, you may find yourself bothered by one or more of the following qualities of technology: ubiquity, amplification (or leveraging), stealth, ephemerality, sovereignty, and dehumanization.

## Ubiquity

Technology is everywhere, massaging us so completely that the human–machine symbiosis has become more a continuum than a partnership.

In an average day we take not a single step that doesn't require us to walk down a path defined by technology. In the preface to *The Evolution of Useful Things: How Everyday Artifacts—from Forks and Pins to Paper Clips and Zippers—Came to Be as They Are,* Henry Petroski describes our situation:

> Other than the sky and some trees, everything I can see from where I now sit is artificial. The desk, books, and computer before me; the chair, rug, and door behind me; the lamp, ceiling, and roof above me; the roads, cars, and buildings outside my window, all have been made by disassembling and reassembling parts of nature. If the truth be told, even the sky has been colored by pollution, and the stand of trees has been oddly shaped to conform to the space allotted by development. Virtually all urban sensual experience has been touched by human hands, and thus the vast majority of us experience the physical, at least, as filtered through the process of design.

Technology is not simply all-pervasive. It is also massively interconnected. It consists of myriad diverse, specialized parts that are fused together in an intricate pattern of interdependency. A god's-eye view of postmodernity reveals an elaborate system of people-to-people, people-to-machine, and machine-to-machine relationships dedicated to the survival of the technological infrastructure that surrounds and nurtures us. As with the biosphere, if one part of the system fails the rest of the system suffers.

Because of its overwhelming pervasiveness and intercon-

*What do machines talk about when they get together?*

nectedness, technology needs to be considered as a secondary
ecosystem in its own right rather than just a whole lot of stuff
that we simply add to the primary natural ecosystem. This secondary ecosystem, the tecosystem, is the human-made part of
reality. It includes all parts of our technology, from salt shakers
to satellites, that share with our natural ecosystem the need for
the connectedness of a highly complex web of systems and
subsystems in order to survive. It is within the tecosystem that
Y2K bugs thrive. It is upon the tecosystem's unfathomable
complexities that they feast.

While technology has been around even longer than the
wheel — early humans had quite sophisticated tool kits — the
deep level and all-pervasiveness of the present connectedness is
a distinctly postmodern event. Through raw materials distribution, worldwide marketing, and the global information processing system needed to keep it all organized, there is no such
thing as an isolated technical event anymore.

It would be nice if we could all agree on certain technologies we don't like (junk mail databases, disco decor, bell-bot-
tom jeans) and say, begone! But the fact is that the very nature
of the tecosystem does not allow selective meddling. Pull on
one thread and the entire tapestry begins to unravel. Life in
the tecosystem is an all-or-nothing affair. Not that you can't
meddle and pull; that is, after all, how change happens. Just
expect a tapestry that looks frayed before it looks better.

That in turn helps fuel our growing sense that we have
lost the future to a world in which technology perpetuates itself
without our guidance or input. And with this feeling comes the
onslaught of what Winner calls the three U's of technological
progress: uncertainty, unpredictability, and uncontrollability.

By far the most intimidating aspect of ubiquity is the fear
of technology's omniscience: databases that know too much

about our buying habits, software programs that watch where we go on the Web, microrecording technology that helps one "friend" trip up another. Even more bothersome than being surrounded by ubiquitous technology is being secretly watched and recorded by it.

*The tecosystem has no private parts.*

### Amplification (or Leveraging)

What is a lever? It is an aid to action that allows human beings to wield an "unnatural" amount of power.

The desire to be more than ourselves has always been with us. In the old days, to paraphrase Joseph Weizenbaum in his book *Computer Power and Human Reason*, levers extended raw muscular power. Humans placed a bar or rollers under a rock to lift or move it when it couldn't be moved directly by human effort.

*An amplifier is a lever made current.*

Some of us are still aware of the leveraging possibilities of a car jack. But today we mostly plug our levers into a wall socket. An amplifier is just an updated version of the lever, designed to take advantage of modern power sources.

Much of the nostalgia that permeates the social critique of technology focuses on those halcyon days of simple tools when life moved along at a leisurely pace and amplification was not the threat it is today. Tools used to be rather subdued in nature, their effects confined to a rather limited sphere of influence.

In contrast, today's machines leverage human action in godlike proportions. Pushing a button, pulling a trigger, or making a phone call can result in an awesome display of the power of the few over the many, both near and far away. We have, to paraphrase Captain James T. Kirk of *Star Trek* fame, the power of the gods without the wisdom.

## Apraxia

*Amplification through technology occurs within the medium of time as well as space, presenting one of the most poignant differences between the ecosystem and the tecosystem. While failure within the ecosystem can take decades or centuries to play out, failure in the tecosystem accelerates events exponentially.*

*We see this in our use of just-in-time inventory systems. For economic reasons, manufacturers order parts on a short-term basis rather than stockpiling them, thereby ensuring paralysis when a strike or failed communication or transportation system halts the flow of supplies.*

*We see it in our hair-trigger nuclear industry. The Soviet Union's Chernobyl disaster, as well as a number of close calls with our own defense systems, reminds us how quickly and lethally events can go awry within complex technological systems.*

*We see it in ourselves when the lights go out. Contemporary history is riddled with examples of how otherwise law-abiding communities degenerate in fairly short order when electrical systems fail.*

*In* Autonomous Technology *Langdon Winner calls this phenomenon apraxia. He borrows the term from the medical world, which defines it as the inability to perform coordinated functioning due to brain lesions. Its appearance in the tecosystem can be bothersome, to say the least.*

From the Greek:
"a"—without;
"praxia"—action.

## Stealth

If you connect to *www2.jun.alaska.edu/edtech/jason/jason-new.jpg,* you will see a picture of me with both hands at my sides, smiling brightly into the camera.

If you then go to *www2.jun.alaska.edu/edtech/jason/jason-old.jpg* and look at the original picture, you will see me with one hand in my pocket, still smiling brightly into the camera but looking a bit sloppy and unprofessional — not quite Web-ready.

The process I used to go from sloppy to preppy involved about a half-hour's work using a digital retouching program. I copied my "good arm," inverted it, superimposed it on my "bad arm," did a bit of doctoring, and presto!—there I am, representing a moment in time that never existed. While I was at it I took care of a few areas on my rock-washed jeans that looked like wet spots—no sense in needlessly embarrassing myself in cyberia.

"Seeing is believing" has been obsolesced.

Of course "seeing is knowing" has never been the adage. Believing what you see has always been an act of faith, our eyes being such a limited technology themselves. But seeing used to be a calculated, educated leap into the unknown based on reasonable assumptions of veracity—assumptions that no longer exist in these digital days.

When I tell my students about the process I used in posting my doctored picture on the Web, I ask them, "Is what I did wrong?" They invariably say no. Then I ask, "Does what I did bother you?" They invariably say yes. It is a unique kind of neurosis in which we are bothered by something that isn't wrong.

Behind the unease is a fear of stealth, the uncomfortable feeling that something has slipped in underneath our radar and penetrated our personal space undetected. We don't like it because we weren't consulted beforehand. Nor were we informed afterward. For many people, life in the infosphere has become a game for only the faithful, the suspicious, the unaware, or the apathetic.

And I am speaking not just about information, which proliferates at a rate that defies the efforts of all of us to verify even a fraction of what we consume on a daily basis. In the infosphere we have come to expect stealth and mass filtering just to make the trillions of bits of chatter among global villagers manageable.

I am also talking about "things." Technology as stuff seems to have snuck up on us, easing its way into our culture too gradually for us to notice. Once it became part of the landscape we were too busy to reflect on it as we raced to keep up with it. The stage 3 and stage 4 literacies, forgetting and seeing plus evaluating, are hard to accommodate simultaneously.

*When something feels good it's hard not to ask for more.*

Railroads begat highways, planes begat rockets, the UNIVAC begat the PC—all based on what Weizenbaum calls the pig principle: if something is good, more of it is better. There was good reason to adopt this view. It was the philosophy underpinning limitless economic growth. What was good for General Motors really did seem good for America.

But if we are willing to look, we are compelled to admit that our "new" technology touches something familiar. In hindsight it is easy to see that the sport utility vehicle didn't really sneak up on us but was the inevitable realization of the wheel, just as the handheld computer was the inevitable realization of the stone and chisel and the waxed writing board used by earlier humans.

It is our lot in life, so it seems, to reconcile ourselves to a present that arrived unannounced.

### Ephemerality

In retrospect history seems to have been kind. We were allowed a few millennia of relative downtime between the Agricultural and Industrial Revolutions, then a few centuries' respite before the Information Age urged us into overdrive.

Now the changes come so quickly we should speak of phases rather than ages: the phases of automation, information, communication, presentation, all within years of each other, each with the potential to transform the tecosystem in powerfully unpredictable ways.

The model for today is life in a constant transition zone, in which change never slows enough to coalesce into anything as solid as an age, in which downsizing replaces steady state as the new metaphor for economic stasis, in which curiosity replaces tradition as the assumption of social consistency. While knowledge knows no limits, it has an ever-decreasing half-life.

We are reminded by McLuhan that our technologies extend us like hardware peripherals: the wheel extends our feet, the telephone our voices, the electric light our eyes. One main purpose of these peripherals is to bring us human beings to a physical level more in line with our imaginary levels.

But we are reminded by the ephemerality of these extensions that at the heart of the relationship between ourselves and our technologies is an emotional attachment.

To overcome our inherent insecurities, many of us spend an inordinate amount of energy trying to limit change. Developing attachments to rapidly evolving technology only amplifies and exacerbates the stress involved in this process.

One of the most difficult challenges we postmodern humans face in a world gone ephemeral is to maintain an unwavering core of values while remaining open to change. We need to be solid but supple, lasting but lithe, fixed but unformed, immutable but innovative, eternal but elastic.

Today's goal is to try to be at peace in the eye of the hurricane and, on a good day, even enjoy the ride.

### Sovereignty

Seeing and being merge with technology in a stage 4 literacy dance because all interactions, whether with a person or a machine, are transactional. We cannot come into contact with anything—mineral, vegetable, animal, or digital—and not become at least a little like the thing itself.

Because a machine is so unyielding in nature and programmed to behave within such tight confines, we submit to it much more than it submits to us. And because tools are just ideas with clothes on, we become not only like the tool but also like the social forces behind it.

*Technology controls us by our own design.*

The thing we think we control because we made it and therefore presume to understand and have dominion over it actually controls us by our unconscious design. It is our sovereign.

It may seem the other way around—that the car offers us untold freedom and control. But there is no driving a car without being driven by it. We mistake our "right" to go slow or fast, to drive the car wherever roads will take us, as freedom, as liberation from the tyranny of locality. After all, as we are informed by James Burke in his book *Connections*, during the Dark Ages the average person wandered no more than seven miles from home.

What we have in fact is a proscribed kind of freedom, with strict behavioral boundaries that make us not the driver but the content of the car. We remain useless until we bend to the limitations it imposes. Should something like Y2K threaten the existence of the tecosystem, we rediscover that we are completely at its mercy.

The net result is the glorification of what Winner and others call reverse engineering. Once a technology exists it is difficult if not impossible not to begin considering life's possibilities in terms of the activities the technology favors.

Also known as technological determinism, reverse engineering becomes part of every imperative we can imagine as the ultimate measure of productivity. We work backward from the means to the end and forward into a relentless cycle of change because of the overwhelming presence of technology predisposed to change.

## Thou Shalt Word Process

*A good example of the reverse-engineering phenomenon is found in word processing, which transforms a symbolizing process formerly reserved for a literary elite.*

*Once upon a time literary elitists felt comfortable engaging in the thoughtful but laborious mechanics of handwriting or typing. With our word-processing programs we have transformed this task into a reactive, egalitarian literacy expected of everyone.*

*During a three-year period in the early 1980s at my university, word processing went from being a possibility to being the only option for any written work. Overnight we had no excuse not to use it.*

*The same will happen with pictures and colored ink. They will be* so easy to incorporate into our documents and messages that there will be no reason not to, whether or not there is an immediate and compelling reason to do so. Reasons, however reasonable, will be developed in hindsight.

Editable words, edible worlds.

*Unlike our parents, we postmodern humans sit at our computers producing and consuming worlds of words on a regular basis. Our constant goal is to make way for the next cycle of fashion or innovation or dissatisfaction that can be temporarily cured with something new.*

## Dehumanization

The most common cry against being overwhelmed by technology is the wail of dehumanization. We come into this world screaming for attention, and we leave it the same way. In between we fret to be heard.

When a machine misspells our name, erases something we mistakenly told it to erase, or — perhaps worse — doesn't understand that a bomb dropped on a city kills people regardless of character, we cry dehumanization.

When the things we place in between us separate us rather than connect us, diminish us rather than improve us, amplify our frailties rather than challenge us to overcome them, we cry dehumanization.

When we rail like this among ourselves in an unmediated environment we are considered immature or selfish. When we rail against our metallic and silicon counterparts we are considered justified because they cannot even hear us to ignore us. With them we don't even get the right to be misunderstood.

Much of technologization occurs because there are simply too many people in the world to organize without the help of machines. It is our response to a world overwhelmed by people, most of whom would prefer the angst of an ordered, mechanically mediated life to the insecurity and opportunities of living in organic chaos.

*Robert Waterman notes that it's almost a law of nature that structured work drives out unstructured.*

Inherent in this preference lies the crux of our dehumanist experience: complexity always overwhelms simplicity. As Winner puts it in *Autonomous Technology,* more highly developed rational-artificial structures tend to overwhelm and replace less well-developed forms of life. In so doing, a technological universe absorbs its non-technological opposition.

### Victimyopia

The net result of ubiquity, amplification, stealth, ephemerality, sovereignty, and dehumanization is a feeling of loss of control.

*We thought we were in control.*

For us inherently emotional and insecure postmodern humans there is no greater assault on the psyche. We become immobilized by distinctly postmodern kinds of anxiety, all centered on who or what is in control of our lives.

We suffer, alas, from victimyopia. Through apathy, resignation, despondency, or lack of vision we believe ourselves to be overwhelmed, at the mercy of, and forever dehumanized by the forces of technology.

So what's to be done? Can't we take a pill and make it all better?

The sensible solution—the one I teach to my technology assessment students and intend to explain in this book—is to gain control over the beast, the force of technological darkness, by attaining stage 4 literacy.

# Two

## See + Evaluate = Control

Marshall McLuhan, the writer who took literary minimalism to new heights by leaving out every other sentence, was above all a mesmerist.

Thus we, his students at the University of Toronto, used to sit listening raptly to the master storyteller, afraid to question him lest we show our ignorance, all the time suspecting he was from another planet more highly developed than ours.

Among his many unique gifts was the ability to "see" a piece of technology: to look backward, for instance, from the photocopying machine to the fall of the Roman Empire, which he blamed on a labor strike among papyrus workers, which caused a shortage of the papyrus needed to keep the Empire's bureaucracy coordinated and in control, which caused communication and other vital networks to disintegrate.

He made us realize that the same thing could happen to us if all our photocopiers ran out of paper. We were connected to the Romans by our mutual need for information, and, not having it, we would fall like the Empire.

See?

For many years it has been my goal to have my educational technology students *see* technology. And the ultimate goal is control, which begins with clear vision, which begets stage 4 literacy, which directs the control—which of course, as many philosophers of technology make quite clear, is illusory if not impossible.

*Better to assume control than to assume you have none.*

Yet I suggest that the only illusion more destructive than believing we can control our technology is believing we can't. So choose your illusion.

The rest of this book assumes that we can bring a sense of balance to the technophilia and technophobia that permeate much of our culture by raising the issue of technology as a force—a kind of beast in the electronic jungle—that, when understood, can be directed. If you share this illusion, read on. If you don't, read on anyway and see how the other half thinks.

## Questioning Technology

What is it that we want ourselves and our children to see about technology?

As mentioned in Chapter One, stage 3 literacy involves forgetting on a deep level. That is how technology thwarts tradition, gadget by gadget, quietly, in ways we can judge only in illuminated hindsight. Stage 4 literacy seeks to reverse this process, to move what is in the background to the foreground, to question the path as we walk on it.

What vantage point best prepares us for carrying out this task? I pose the issue of assessing technology to my students in the following way.

Suppose there existed a Science and Technology Administration (STA) which, in the manner of the U.S. Food and Drug

Administration (FDA), was charged with assessing not foods and drugs but the possible impacts of a technology before it was released into mainstream society. Then suppose you were an investigator for the STA. What questions would you ask about a technology as you sought to understand its potential impacts?

The questions posed by a conscientious STA investigator would fall into three basic categories that correspond to the next three sections in this chapter.

"The Technology Itself" section examines the characteristics of the technology in semi-isolation. Using the mechanist's religion of reductionism, the STA investigator asks questions that help us understand the unique characteristics and properties which a technology possesses.

"The Goals of the Technology" comprises questions that explore how technology is imbued with purpose and manipulated toward particular ends.

"The Effects of the Technology" zooms out for a holistic examination of questions concerning technological impacts on three areas of life: the environment, ourselves, and our institutions.

## The Technology Itself

In *Sanity, Madness, and the Family* R.D. Laing observed that it was pointless to try to fully understand insane persons apart from their environment because they *were* their environment. Their mental state was just one inextricable strand of a tapestry of collaborative neurosis they had helped to weave. And, quite possibly, it was actually the sanest reaction to their situation. For this reason Laing recommended that psychiatrists use their couches cautiously, cognizant of the fact that a full

*Margaret Mead admonishes us never to doubt that a small group of committed citizens can change the world. It's the only thing that ever has.*

*Marshall McLuhan: We wear each other. Jean-Paul Sartre: We wear on each other.*

understanding of a condition demanded observing the patient in the whole context of the natural social environment.

It is no different in the field of physiology. Tissue observed in a petri dish yields certain clues. Observed in the greater context of the human body, it yields others. Used together, the two approaches form a powerful alliance of knowledge.

In much the same way there is really no such thing as understanding "the technology itself." Technology is a symptomatic metaphor, inseparable from the forces behind its creation and interpretation. And, quite possibly, it is a rational reaction to the irrational quest of *Homo sapiens* for a better life through, to paraphrase Carl Sagan, exosomatic enhancement.

Like tissue, technology can be observed under a microscope as well as in the whole, yielding different kinds of information which, taken together, draw a powerful and detailed picture. Therefore as we attempt to see the technology itself we do so knowing we have caught something temporarily in our grasp, a tranquilized animal, a beast that we tag and study before returning it to the wild.

*Reductionism vs. holism: some of the parts vs. the sum of the parts.*

We view particular technologies the way scientists view physical events — as isolated phenomena — and seek above all to understand how a technology behaves and how we interact with it on a functional level. In so doing we are better able to discern its impact on ourselves, our institutions, and the environment when it is returned to its natural habitat: the holistic world of human events.

To help us see and evaluate the technology itself — always remembering that the ultimate goal of seeing and evaluating it is controlling it — we, as STA investigators, need to ask a number of questions.

**What is the technology, who uses it, and what does it do?**

## SEE

The goal in considering this question is to obliterate technology as background, to bring it into high relief against a backdrop of the habitual.

Quite often I do exactly that with my students. I place a common object on a white sheet of paper while the classroom is dark, then turn the lights back on — and the ordinary makes a dramatic entrance. We then analyze the object, with which most of us have shared an ordinary but personal history for many years, as though we were seeing it clearly for the first time.

You can do it yourself. Be an anthropologist examining the technology of your own life. Think as much as possible in clear, objective, and precise terms about it.

*A car is a car until it turns into a driveway.*

*— Boy's Life magazine, some time ago*

Focus first on the function of a particular tool or machine. What are the many ways you and others have come to depend on it? If you were from another planet what would you make of this toaster? This ballpoint pen?

That's an obvious but not trivial place to start.

---

### A Rose Is a Rose

*My middle school football coach used to begin each season by clutching a football in his outstretched hand and solemnly declaring, "Gentlemen, this is a football."*

*In the same spirit, the assessment of an automobile begins with the declaration that this is a car.*

*And it continues by stating who uses it and what they do with it. The principal function of a car may be to transport people faster than they can walk, but it has many other functions: bedroom, attack vehicle, mobile office.*

We ask:

- ▸ Who commonly uses the technology and what for?
- ▸ What is the common experience or understanding of the technology?

**What is the technology made of and how does it work?**

### SEE

*How in the world, Ivan Amato asks in amazement, did humanity learn to transform the raw stuff of the wilderness into the contemporary zoo of materials?*

The tecosystem, you will recall, is the ever-evolving infrastructure of interconnected technologies that circumscribe our behaviors on every level, from the physical to the philosophical.

The inverse of the tecosystem is the inner workings of a technology, a miniature universe comprising materials, connective subsystems, even instructions about its origins and operation, the interdependence of which gives it a kind of inanimate forward momentum. As noted, there is no such thing as "the technology itself." Nor is there in our postmodern world a single technology as a monolithic construct.

---

### At First Sight

*To appreciate how the intra-tecosystem of a technology works, try an exercise I conduct with my students.*

*Get up and leave the room you are in. Then, before reentering, adopt the following mindset:*

*You have arrived from another time, place, or planet. In your*     *My-clue-in.*
*mind you are wearing a white lab coat and holding a clipboard, the ever-vigilant notetaking anthropologist who sees everything as a clue.*

*As you walk back into the room, see it as a technological system with a number of interconnected subsystems. Let the medium become the message, context become content, background become foreground. See it for the first time.*

Just as you can work backward from how you live to the belief system that explains your actions, you can also work backward from a technical structure to understand much about the people who created it, from their materials to their mythologies. Look about and be alert. The STA is counting on us.

## + EVALUATE

We ask:

- ◪ What does the technology or technological structure do, and what is it made of?

- ◪ What subsystems are present, and how do they interconnect?

- ◪ What can we assume about the evolution of the people who created the technology? About their understanding of science? Their mythology? Their manual dexterity?

- ◪ Do we see any implicit social structure in the way the technology is set up? Is there a hierarchy? A lowerarchy?

- ◪ Is there writing anywhere? If so, what does it say, literally and figuratively?

- ◪ Are there pictures anywhere? If so, what are they telling us?

- ◪ Do we see indications of mass manufacturing, individual creativity, or both?

- ◪ What energy sources does the technology require, and where do they come from?

*Genius means little more than the faculty of perceiving in an unhabitual way, writes William James in his* Principles of Psychology.

### Is the technology more a tool or a machine?

## SEE

We can make a number of significant distinctions between tools and machines that help us understand who we are in relation to our artifacts. Consider some differences between them.

- ◪ Tools adapt to the rhythms of the human user, while machines force us to adapt to theirs.

- ◪ Machines exhibit an inherent regularity of behavior, while tools exhibit little or none.

- ◪ Tools are almost entirely dependent on human energy to operate, while machines use power supplied by non-human sources.

- ◪ Machines usually have some built-in intelligence (toasters know when to pop up), while tools depend on the human user to supply most or all of its operational acumen (a chisel awaits its master's grip).

The reality is that few pieces of technology are wholly tool or wholly machine. They are combinations, points on a tool-to-machine continuum for each of the characteristics just described. It is up to us as STA investigators to determine what best describes a technology's unique blend of its toolness and its machine-like qualities. Doing so helps us see the technology with a clarity and vividness that may be quite new to us.

The functionality of tools and machines can also be understood as an expression of the societies in which they have emerged.

*New technologies, says poet Wendell Berry, should be repairable as near to home as possible by people of average intelligence.*

We generally associate machines with the coming of the Industrial Revolution and tools as instruments of the preindustrial artisan. Implied in this distinction is a profound confusion in our sense of place. We often think of tools as locally made, maintained, and improvised and machines as products of a manufacturing process carried out by strangers in faraway places using foreign materials and unknown techniques. Our increased dependence on machines over tools has decreased local authority and personal autonomy and begotten the global economy.

Placing the technology on the tool-machine continuum can help us see it more clearly. And lest we forget, seeing is the first step toward control.

### + EVALUATE

We ask:

- How much does the technology make or allow the user to do?
- Does the technology exhibit a regular, predictable behavior?
- Is the technology locally made?
- Can the average person fix the technology, or does it require a specialist to run and maintain it?
- How much intelligence does the technology have?

### Is it tool or technopoly?

### SEE

We can cast the functionality of tools and machines into a wider and deeper arena by considering them as expressions of mythology.

The centerpiece discussion in Neil Postman's *Technopoly: The Surrender of Culture to Technology* is the issue of who in the drama of human evolution gets to play the role of god: we or our inventions? As we postmoderns have discovered, when this kind of question confuses us it is often more expedient to upgrade our mythology than to modify our technology. And so we ask our gods to serve in the employ of our machines.

That response is what Postman calls technopoly. He argues that getting to technopoly was a three-stage process.

The first stage encompassed the days of yore, in which cultures invented tools to support their belief systems. Eventually someone invented the water-powered alarm clock to get the faithful to church on time; people developed hand tools to craft bigger and better cathedrals and castles; they crafted writing instruments primarily to serve the bureaucracies of king and church. Postman sums up this period of artifactual human by stating that "tools did not attack . . . the dignity and integrity of the culture into which they were introduced." Belief systems, in fact, "directed the invention of tools and limited the use to which they were put."

The second stage was a period of "technocracy" in which invention and belief were separated, allowing the pursuit of science to flourish unfettered by the strictures of scripture or general ethics. During this period in the West, the Roman Catholic Church presented the natural world to its faithful as something to be conquered, opening the door to what Mother Nature might consider unnecessary and invasive exploratory surgery. This was the period of Copernicus, Kepler, and Galileo, Newton, Bacon, and Descartes. It was an age in which knowing was next to but apart from godliness and belief was the antithesis of understanding. Postman notes that theology, "once the queen of all sciences, was reduced to the status of court jester."

Technopoly enters into the third stage as invention directs people's belief systems. That is, we engage in reverse engineering of mythological creation, working backward from our inventions to a belief system that supports their existence. Questions of why become fully subsumed by questions of how. As Langdon Winner notes in *Autonomous Technology*, efficiency becomes its own closed system of worship.

*Nostalgia:*
*Do yore thing.*

*Albert Borgmann notes that in a traditional culture, a means is always inextricably woven into a context of ends.*

*The can-do attitude: Because we can, we do.*

Postman paraphrases Alfred North Whitehead by noting that the greatest invention of all during this period was invention itself. Invention, scientific management, the triumph of information over insight, the dominion of technology over humanity: these are the hallmarks of a technopoly.

Seeing technology aright thus involves finding out where it falls on the tool-to-technopoly continuum.

*Technianity:*
*Belief systems*
*created of, by, and*
*for machines.*

What's next after technopoly? Technianity: the creation of belief systems by machines for machines. As artificial intelligence and self-manufacturing evolve and combine, technianity seems inevitable. Our machines will begin to exhibit the same impatience with us that we exhibit with them.

### + EVALUATE

We ask:

- Does the technology offend, interrupt, or challenge the local mythology or status quo?

- Is there implicit in the invention of the technology a consideration of ethics, community values, or spiritual beliefs?

- Is the technology developed within the closed system of efficiency or in a self-fulfilling system of knowledge for knowledge's sake?

**Is the technology more an information machine or a work-producing machine?**

### SEE

As STA investigators we need to understand that the distinction between an information machine and a work-producing machine draws attention to the fact that technology of the

Information Age — "thinking" technology — is quite different from the Industrial Age's "doing" technology.

Industrial Age machines are those that produce heat or work and often have limited decision-making capabilities — like toasters. Their actions are more localized and less leveraged than the actions of Information Age machines, which do a great deal of thinking, projecting, and connecting over time and space.

The ability of information technology to engage in decision processing — the forerunner of thinking — gives rise to the spookier side of sci-fi, the side that leaves us feeling that our machines are talking about us behind our backs. We see in them the specter of dehumanized automation, loss of privacy, and centralization of power over the many.

*When all is said and done, Ernst Braun writes, all human activity is and always has been based on information — indeed, any action other than random motion is informed.*

## Where Was I?

*It is our information machines' ability to create a semblance of immediacy that confuses our sense of place.*

*As Joshua Meyrowitz notes in* No Sense of Place: The Impact of Electronic Media on Social Behavior, *"Electronic media destroy the specialness of place and time. . . . Through such media, what is happening almost anywhere can be happening wherever we are. Yet when we are everywhere, we are also no place in particular."*

*On a very practical level, Meyrowitz writes, new media "merge many formerly distinct social situations, blur the dividing line between public and private behaviors, and tear apart the once taken-for-granted bond between physical position and social position." That is, we find ourselves with no direct connection between who we are and where we are.*

### + EVALUATE

We ask:

- Does the technology deal in information rather than work?

- ◿ Can it think? How smart is it?
- ◿ Does it aid and abet automation? Centralization?
- ◿ Does it encourage or discourage autonomy and privacy?

**How does the technology extend people's senses and capabilities?**

### SEE

Every tool and every machine, like a peripheral attached to a human warmware docking station, extends our sensory apparatus, physical capabilities, intelligence, or emotional experience of reality. This extension gives technology an innate kind of power that makes human action dangerous in the wrong hands, saintly in the right ones.

That is why we, as STA investigators, need to see clearly how a technology extends human capability.

Deity and demon both embrace a multitude of technologies. These technologies seem to be a part of us, not so much challenging our power as adding to it. Our attachment to them, both physical and emotional, makes it important that we understand how this extension manifests itself.

Extension can often be quantified on a ballpark basis. The car that carries 1,000 pounds of goods extends my body, which can carry 100 pounds, by a factor of 10. And whereas the car can easily go 1,000 miles and more, I could probably carry those 100 pounds only about 100 feet before collapsing and begging someone to call 911.

Most Industrial Age technologies are easily understood in these terms. New technologies, which extend more intangible qualities such as mind and memory, are a bit trickier.

Yet it is still feasible in some ways to compare life before and after the adoption of an information technology. In some cases we can compare calculations per second as well as the time taken to recover and coordinate information. Or we can simply list activities not possible before the technology was introduced. Wherever helpful and feasible we STA investigators should try to quantify the difference a technology will make.

McLuhan, the popularizer of this concept of extension, viewed people as two-dimensional mental constructs without interiors. His analysis of extension proceeded from the belief that people gain knowledge, the raw material necessary to create meaning from experience, primarily by means of the five senses. Jerry Mander, in *Four Arguments for the Elimination of Television,* adds to the five senses four internal senses or capabilities that can also be extended: instinct, intuition, feeling, and thought.

To McLuhan, TV extends seeing, hearing, and community through shared sensory experience. To Mander, TV projects a prepackaged emotional response and sense of perspective that dominates the unconscious of the consumer.

To McLuhan, TV "retribalizes" us. To Mander, that is just a euphemism for the standardization needed by advertisers and mass marketers to be successful in associating interior senses of complacency with exterior objects of pleasure.

McLuhan sees TV-absorbed humankind returning to the village—one of global proportions. For Mander it is blasphemy to call it a village; TV is just a sellathon in which corporations with commercial interests define community standards and run the town meetings.

## Automoselves

*The automobile extends our feet by increasing our speed.*
  *It extends our eyes with headlights that allow us to see at night.*
  *It extends our ears through radio and antenna.*
  *It extends our backs by increasing the load we can transport.*
  *It extends our voices through the horn.*

### + EVALUATE

We ask:

*The bad news, Donald Norman warns, is that technology can make us stupid too.*

- Which of the five external senses does the technology extend?
- Which of the four interior senses does it extend?
- Using a ballpark estimate, by how much are these senses extended?

### How does the technology limit people's senses and capabilities?

### SEE

The opposite, according to McLuhan and others, is also true: Whenever a technology amplifies, it also weakens. Whenever it connects somewhere, it disconnects somewhere else.

*To know what a car won't let you be, take a train. Or, better yet, take a hike.*

Because we are overtaken by the connective properties of a technology (in truth, we probably bought the technology because of such properties), we don't focus on how it reduces or disconnects us from who we are. No one wants to be a party pooper. Why focus on the fact that a microwave oven deconstructs and scatters the family by making instant, noncollaborative cooking abundant when you can revel in the freedom it brings family members to take advantage of individual opportunities offered by hot food on demand?

We can place every technology on scales that allow us to weigh not only what we gain by adopting it (the connect factor) but also what we lose (the disconnect factor). Understanding how technology limits and extends our senses and abilities is a solid place to start.

---

### Dysautomoselves

*Amending our analysis of the automobile (see "Automoselves" on page 57), we may observe that:*

*The automobile not only extends our feet by increasing human speed but also amputates them by keeping them still and out of contact with the earth. The faster we go, the less detail we can observe.*

*It not only extends our eyes by increasing our night vision but also atrophies the muscles we need to see in the dark.*

*It not only extends our ears with radio and antenna so we can tune into the national debate but also disconnects us from over-the-fence chatter in our own neighborhood.*

*It not only extends our backs by increasing the load we can transport but also reduces the muscular integrity we need to carry even a lightweight package.*

*It not only extends our voices through the horn but also decreases articulation to a beep.*

---

### + EVALUATE

We ask:

- Which of the five external senses and four interior senses does the technology limit and in what ways?

- Can the limits be quantified?

- How does the technology reduce abilities we might ordinarily use?

How does the technology disconnect us from opportunities that might otherwise be available?

## How does the technology alter temporal and spatial relationships?

### SEE

In earlier days, when few people wandered far from home and time unfolded more fluidly, our forerunners regarded time and space as superior beings or gods to be respected and feared. But with the rise of modern and postmodern humans, time and space have decreased in stature just as nature has—from supernaturals inspiring worship to peers with whom we coexist and finally to nuisances that govern much of the tedium of our lives.

The advent of powerful transportation and communication technologies has greatly amplified this fall from grace. In fact, people created these technologies specifically to overcome time and space. And the process has helped cultivate our depreciation of these two facets of the universe as design flaws in nature. We hit the wall of time and space and reach for a technology to circumvent their limitations.

All technology somehow rearranges our experience with time and space, contracting it, expanding it, recasting it, and in some cases obliterating it entirely.

Mechanical travel does more than compact space. It amplifies our attempts to use technology to bend time and space to our will. It allows us to live in a state of suspended now, from end point to end point.

McLuhan noted that a plane trip is little more than a horizontal elevator ride, an event with a beginning and end but no

*Sven Birkerts observes that we now find it as unthinkable to walk five miles to visit a friend as it was once unthinkable to speak across that distance through a wire.*

## Shapeshifters

*Time and space are the most indestructible elements of the natural world, yet they move through our lives pliably, like shapeshifters metamorphosing from one incarnation to another.*

*A microwave oven shrinks the amount of time we spend cooking but expands the time available to skateboard, watch television, or shop.*

*An airplane contracts the space between two distant places but expands the world space available to us.*

*A telephone obliterates the space between us and the persons we call but solidifies the physical distance between us and our friends.*

*An elevator compresses space for the gravitationally challenged among us while allowing them to attain new heights.*

*Home exercise equipment shrinks the time we spend traveling to the gym but expands the time we spend on our stationary bikes.*

*When we cook too quickly we miss the slow evolution of aroma.*

middle. The windowless ride is the operant metaphor for the accelerated pace of postmodernity. As we speed up, what becomes all-important is not exploring and enjoying the path we travel but overcoming it, as if it were an obstacle in our way.

The relative clumsiness of physical movement compared with the speed and dexterity of ideas probably made the development of improved communication technology inevitable. Our communication tools and machines represent an attempt to eliminate the suspended state of now and to correct the perceived tyranny of time and space that separates us from the distributed opportunities of the digital era. They do it, as Nicholas Negroponte, author of *Being Digital,* would say, by moving bits rather than atoms.

*Moving information, not people — a semblance rather than assemblance.*

The result of this indirect method of travel is that the relationship between the space we physically occupy and our experiences of life becomes even less clear than when we commute between home and work. "To have information of an environment is to be partially 'in' that environment," Meyrowitz writes in *No Sense of Place.*

---

## You Can't Go Home Again

*A confused sense of place begins passively—to watch TV is to be displaced from where we sit and stare.*

*The confusion becomes even more amplified as we go interactive. When our atoms travel we leave one place and enter another. But when our bits travel we occupy two places at once.*

*Part of traveling in bits involves filtering: the deliberate attempt to serve bandwidth limitations and, often, to purposely reconstitute content. This happens technically through camera angles and the compression of video bits to speed up transmission.*

*It happens editorially through omission and inattention to diverse viewpoints.*

*And it happens commercially by allowing advertisers' interests to determine content. But what is most important is that it happens.*

*Confusion and displacement border on alienation. Where is home in a distributed community?*

*Kids love it. Many adults turn ashen-faced.*

---

Efficiency—the overcoming of time and space as effectively as possible—came into its own in the Industrial Age, when rational thought went external through the creation of machinery. Now, in the Information Age, communication technology realizes a kind of superefficiency by transporting the representation of experience instead of the experience itself.

*Stone Age people, Jerry Mander speculates, had more than twice the leisure time we do and used it to pursue spiritual matters, personal relationships, and pleasure.*

Think about how much of our day we spend supporting the technological infrastructure, whether we're working to make car payments, comparison shopping for the best VCR, traveling up another late-night software learning curve, or simply fixing the screen door that won't close properly. If free time is what we're after, isn't it possible that the less technology we include in our lives the more time we will have?

Such thinking is heresy, of course, for us civilized post-moderns who prefer the comforts we find in our technological bondage to the brutish simplicity of reality in the raw.

## + EVALUATE

We ask:

- Does the technology speed things up or slow things down?

- What experiences do we miss or gain in the process?

- Does the technology increase or decrease efficiency?

- Does the technology separate the place of participation from the location of action?

*The Web teaches us to think two places at once. I link, therefore I am.*

- Is our involvement passive or interactive?

- How much of the experience happens in our immediate environment? How much is projected from elsewhere?

- How much is bit travel versus atom travel? To what extent is bit travel filtered?

- To what extent does our time-saving technology cost us time, money, and frustration that was not part of our lives before we adopted it?

## What is the technology's range?

## SEE

*Home on the spatial range, where urban sprawl and com-muter crawl play.*

Range is the extent of a technology's influence in both space and time. A megaphone has a greater spatial range than the human voice but less than a loudspeaker system at a rock concert. Some baked goods, with a rumored shelf-life of twenty years, have a greater temporal range than untreated cardboard left out in the rain but less than the pyramids of Egypt.

We seek to understand a technology's range because it helps us understand the extent to which the technology acts symbiotically to rearrange the environment in which it exists. Just think how motorized vehicles, among other technologies, have rearranged the American continent.

---

## A Day's Difference

*Technology and ruling at a distance developed symbiotically. But so did technology and the spread of democracy.*

*Most of us live in an amplified center-margin spatial range, to use the terminology of Lewis Mumford in* Technics and Civilization. *In this range, transportation and communication technology facilitate the extension of power from the city to its outlying regions.*

*That is in contrast to the ideal Greek democratic city state, whose size, as James Cary points out in* Critical Connections: Communication for the Future, *was small enough to allow any citizen to travel on foot to the most remote point within it and return in one day.*

---

Combining technologies greatly increases range. The symbiotic relationship of papyrus and roads, according to McLuhan in *Understanding Media: The Extensions of Man,* gave rise to nation states as bureaucrats gained the foot-powered bandwidth they needed to distribute and enforce standardized laws and cultural norms. That's why a papyrus workers' strike was able to critically wound the Roman Empire, shutting down its nervous system.

And that should make us pause to consider what might happen if the United Satellite Workers and the Federation of Silicon Chip Makers walked off the job simultaneously.

Electronics, nanotechnology, and nuclear power provide examples of how technology works in a domain that eludes

*For my birthday I got a humidifier and a dehumidifier. I put them in the same room and let them fight it out.*

—Comedian Steven Wright

our direct observation. In that domain the infinitesimally small has mutated into the very amorphous.

Nowhere is this effect more apparent than on the World Wide Web. Its knowbots, no larger than a tiny amount of computer code, roam the Internet looking for answers to our questions: How are my stocks doing today? Are there any new cures for the cancer I might develop? Who wants to be my friend?

Knowbots run invisibly in the background, yet they travel many more miles in a minute than most of us do in a lifetime.

Range over time refers to the technology's ability to endure. One end of the spectrum is characterized by a permanency that makes us feel that in adopting a technology we're crossing a point of no return. The possibility of de-technologizing, even if we had an intense desire to do so, is problematic.

*It is sobering, as Elting Morison remarks, to consider that the Roman Colosseum, built to fill the same needs as Shea Stadium, has lasted two thousand years.*

Technology has a way of ingratiating itself into our existence so thoroughly that we let it hang around, like the guest who comes to dinner and forgets to leave.

Once a factory is constructed or a satellite dish installed, once we adapt to the rhythm of a microwave oven or the opportunities of commuting by car, reverse engineering sets in and we seek a mission to match the technical capabilities at hand. Winner argues in *Autonomous Technology* that it is this kind of reverse engineering that largely determines the *de facto* kind of relationship we have with our technology.

One buzz phrase among planners is "hardware optimization." That is, once an entity purchases a technology, the only way to amortize it is to use it. And the more ways we can invent to use it, the greater the optimization we will realize, whether or not the uses are truly useful.

On the other end of the temporal range spectrum is the planned obsolescence of most consumer tools and machines.

Everything from my Walkman to my washing machine is built to break, shortening the life cycle of the machine and ensuring the steady consumption and production patterns that keep our postmodern economic system humming.

*The only thing that will never escape obsolescence is obsolescence itself.*

In the middle of the temporal spectrum—or perhaps off to one side in a category of dysfunctionality truly unique to postmodernity—are computers and related information machines that become obsolete not because they are broken but because they are lethargic or incompatible with other computers.

I am compelled to toss out my computer every few years because it is incapable of running software programs that I can't help myself from wanting or programs that everyone else is using, thereby putting me in a blue funk or making it and me functionally illiterate.

With any kind of luck my car will last at least a decade. I may own my claw hammer, stereo, and mountain bike for the rest of my life. But my computer lives on borrowed time from the day I purchase it. That is the bargain we make with the computer moguls: give me more power, and I will discard something that is still quite useful.

## Getting Peeled

*Obsolescence comes in many forms. I still feel the visceral reaction I had thirty years ago when I read about a tool designer's sneaky scheme in Vance Packard's* Waste Makers.

*According to Packard, whose work helped pave the way for consumer advocates such as Ralph Nader, the designer suggested that the manufacturer color the peelers to resemble potato peels so users would inadvertently toss them out with the peels and thus buy more peelers.*

*Perfectly good peelers would be obsolesced by the users' sleight-of-hand without their even knowing it.*

## + EVALUATE

We ask:

- ◪ What is the domain or extent of the technology's influence?

- ◪ How far or how short a distance can the technology convey information or goods?

- ◪ How temporary or permanent is the technology? Can we get rid of it if we want to?

- ◪ How does the technology meet its demise? Breakage? Trickery? Incompatibility?

## What is the technology's capacity?

### SEE

In the world of transportation technology we measure capacity in terms of maximum load, axle weight, and cubic storage area units. A car has more capacity than Ms. Gulch's bicycle basket but less than a double-decker bus.

In the world of information technology we measure capacity in bandwidth and throughput, both of which are considerations of the amount of information transmitted per given unit of time. A phone line has less capacity than a fiber optic cable but more than the string connecting two tin cans.

We can view other technologies, from hunting knives to microwave ovens, in terms of the variety of tasks they accommodate. Generally the greater the capacity the more powerful the technology. Wherever we see a concentration of power we should suspect the presence of leveraged action.

## + EVALUATE

We ask:

*In 1997 it took five square feet of physical space to store one terabyte of data. In 1948 it took a space the size of Argentina.*

*—Information Age rumor*

- How much stuff can the technology carry or accommodate?
- How much information can the technology convey?

## How hot or cool is the technology, McLuhanistically speaking?

### SEE

One of Marshall McLuhan's most controversial theories states that the temperature of a technology, in metaphorical terms, depends on the degree of involvement it demands from the user.

*Sometimes I think we forget that something that's cool maybe isn't so hot.*

— June Cleaver

A "cool" medium demands our active participation to make it fully functional. A "hot" medium either does not engage us directly or is so rich in information that it raises at least one of our five senses to such a high degree that our involvement appears more like resignation than participation.

The radio, which plays in the background or foreground whether or not anyone is listening, is a hot medium — so the theory goes — while the phone is cool because conversation demands our attention.

The inherently fuzzy resolution of television makes it an involving, cool technology because we are forced to complete the picture it sprays on the back of our retina — according to McLuhan, we literally connect the dots within our brain. But

*McLuhan's ideas are hotly contested by a number of cool people.*

movies, with far better audio and a brilliant resolution many times the contrast ratio of TV, are hot because we can simply sit in the theater and let the sound and sight overtake us in an act of effortless surrender. The result, in McLuhan's way of thinking, is that the TV watcher participates more raptly than the moviegoer does.

McLuhan's assertions seem to contradict ordinary life. All

of us have experienced moments of listening very intently to the radio, only partially concentrating on a phone conversation, using television as background, and being extremely involved with a movie.

In truth, most of what McLuhan had to say in this regard is tenuous at best. In the case of movies he would probably argue, were he alive to do so, that watching a film is less physically involving than watching TV though it may be equally emotionally involving. It was McLuhan's particular weakness that he didn't concern himself with such things as emotional involvement.

What we as STA investigators can salvage from McLuhan's hot/cool theory is the important concept of involvement: one significant measure of a technology is the degree to which users need to invest themselves in the experience in order for it to be successful.

## McFreud

*In "What If He's Right?" (in* The Pump House Gang*), Tom Wolfe deconstructs Marshall McLuhan and Sigmund Freud by comparing the two giants of modern thought.*

*Wolfe notes that both thinkers opened new doors to understanding the human condition. Both developed revolutionary principles of analysis necessary for understanding diverse phenomena of their modern era.*

*But, Wolfe concludes perspicaciously, both then proceeded to use their work in ways that limited its utility—and annoyed the average person no end.*

## + EVALUATE

We ask:

- How involving is the technology? Does it force us to concentrate on the experience or is it background noise?

### How dependable is the technology?

**SEE**

We assume that entropy and human error will always conspire to render technology fallible. As STA investigators we want to know how susceptible a technology is to these forces.

*Do you deserve a break today?*

Of particular importance in this regard is an issue I raised in considering tools versus machines: when something breaks, who can fix it? If you or someone relatively accessible to you cannot repair it, then the technology is, as James Burke puts it in *Connections*, a trap that has lulled you into dependence while giving you no way out when it fails.

**+ EVALUATE**

We ask:

- When the tool or machine breaks, can we or someone we know fix it?
- Does the technology contain backup systems that offer protection if it fails?

## The Goals of the Technology

Before the tranquilizer wears off and our wild animal escapes into the thickets of the electronic jungle, there is another aspect of its character that we can examine in quasi-isolation. I call it the goals of the technology.

Whereas the preceding section concerned the hows of technology, this section concerns the whys. Again we focus on essential questions in our attempt to see and evaluate in order to control.

# What are the inherent biases of the technology?

## SEE

No living social critic would be caught dead suggesting that technology is neutral — that is, independent of any innate bias or direction. To McLuhan, who considered technology so imbued with bias that it actually became its own content, the idea of technological neutrality was sheer baloney. Technology not only determined but also invented behavior.

While media critics in McLuhan's day (he died in 1980) worried mainly about what people were watching on television, McLuhan's concern was the act of watching. He saw that people were engaged in a ritualistic mass behavior that combined the senses in new ways reminiscent of tribal behavior. And, realizing that technology makes us do things that we otherwise might not do, he declared that the medium is the message. In so doing he helped usher in the era of technological determinism in which we now find ourselves.

McLuhan's theories are open to question. But his belief in the nonneutrality of technology — that the medium is the message — is widely shared. The statements of sociotechnical theorists on this topic are blunt: Kirkpatrick Sale, *Rebels against the Future:* "Technologies are never neutral, and some are hurtful"; Neil Postman, *The End of Education:* "Every technology has a philosophy"; Jerry Mander, *In the Absence of the Sacred*: "The idea that technology is neutral is itself not neutral"; Langdon Winner, *Autonomous Technology:* "Artifacts have politics."

Of course technology is not neutral. Nothing in life is neutral. We are not neutral, the ideas and feelings we have are not neutral, and therefore the things we say, do, and make are not neutral.

And yet you may still have nagging doubts here. Isn't a

hammer relatively undefined until we pick it up and do something with it? Given that a computer can be used to do many things, isn't it more or less neutral until we apply it to some task?

Even those ancient technological inventions, books of wisdom such as the I Ching, or Book of Changes—aren't they neutral when no one is reading or consulting them? As the earliest commentary on the Chinese oracle states, in the Richard Wilhelm/Cary F. Baynes translation, "The Changes have no consciousness, no action; they are quiescent and do not move. But if they are stimulated they penetrate all situations under heaven."

This quandary—the nonneutrality of technology versus its apparent impartiality—dominates the gray area at the heart of many social debates. The argument between the National Rifle Association and gun control advocates provides an excellent example of how this quandary plays out in real terms.

On one hand, those who accept the NRA's philosophy see a gun as harmless until it is touched by human hands. As one NRA radio interviewee said, and I paraphrase, "When a gun can hop out of the box and shoot someone on its own, then I will push for gun control." Gun control advocates, on the other hand, believe that a gun has so much inherent bias toward harmful behavior—the power to threaten, maim, and kill—that we should not trust free will or acculturation to produce only people who will use guns in society's best interests.

Both sides have perfectly valid points that are not mutually exclusive. NRA members wouldn't argue that guns don't threaten, maim, or kill; in fact, that is why they keep handguns in their homes. And gun control advocates wouldn't argue that guns kill on their own; in fact, it is the human user that concerns them. But NRA members focus on the gun as message or content, whereas control advocates focus on the gun as medium or technology. The two sides are led to contradictory conclusions based on their different approaches: the free will of the messenger versus the inherent limitations of the medium.

Still confused? Let's consider the hammer. What *can't* we do with it—or at least what would we be highly unlikely to attempt to do with it? Paint, cook, turn a screw, send email, be gentle and caressing, play kickball, plant corn, to name a few activities. The hammer compels us to engage in a very limited number of actions: hitting and prying. Even these acts are subject to further refinement: hitting and prying things of a certain consistency, size, and resistance.

Now couple this inherent bias toward particular actions with the concept of technology as an extension of human capability. That is, see the hammer as part of your hand rather than a separate "thing" which you hold. See yourself not as a person with a hammer but as a hammerperson, compelled to hit and pry as much as to see or hear because you are "built" to do so. What will you do? You will hit and pry, fully realizing your own intent—along with the hammer's.

While you as a hammerperson are only doing what comes naturally, the intent of a technology sometimes is deliberately engineered. Such was the case, as Winner tells us in *The Whale and the Reactor: A Search for Limits in an Age of High Technology,* when Robert Moses designed some of New York's bridges several decades ago. Moses made sure that the bridges would keep buses, patronized mainly by the lower classes, from reaching the exclusive Jones Beach recreation area. In this way he effectively reserved the beach for the car-owning elite.

Much technological "intent" happens less deliberately than Moses' attempt to keep the hoi polloi out of Jones Beach, decreasing the likelihood we will see it and increasing its power of persuasion.

Take the car as an example. An elementary list of the car's aspects and the behaviors they encourage might include the following:

- Burns gas, which encourages us to go to war in the Middle East, which in turn spurs military readiness.

- Goes more than sixty miles an hour, which encourages us to seek employment away from home, which in turn spurs the creation of suburbs.

- Has headlights, which encourages night travel and nightlife away from the home, which in turns spurs the entertainment industry.

- Goes only on roads, which encourages us to construct highways, which in turn spurs asphalt production.

In the car as in other kinds of technology, every bias toward action begins a domino effect and is itself part of the domino effect begun by another invention.

Another consideration falls under the rubric of implicit bias. It forces us to abandon the closed system of "the technology itself" under which we have conducted parts of our investigation and to look at the wider arena of human events.

This view sometimes is referred to as the politics of technology. According to Winner in *The Whale and the Reactor*, at the heart of this issue is the way in which "choices about technology have important consequences for the form and quality of human associations."

A continuum of human invention that often emerges in this regard is delineated by solar power on one end and nuclear power on the other. Solar power encourages democratic,

*Anything conceived by us contains our bias.*

Technology is a
wayward creature,
according to Ernst
Braun. It's part of
society, yet much of
its activity seems to
be concerned only
with itself.

decentralized, small-scale, locally controllable power produc-tion, the technical aspects of which are within the engineering reach of average people. You and I could research, set up, use, and maintain solar systems capable of providing energy for our families given the determination to do so. Nuclear power, by contrast, is inherently authoritarian, large-scale, and central-ized, requiring a technical elite to create and manage it. Such a system offers none of the egalitarian promise of solar energy and much of the exclusivity of impenetrable, monolithic corporations.

The point here is that technological systems are often inseparable from, in Winner's words, the "creation and main-tenance of certain social conditions" as their operating envi-ronment. Certain technologies need certain political climates and corporate conditions in order to flourish. Because of that, the adoption of technology and particular forms of social order often becomes a package deal. *(continued on p. 76)*

## Built-in Media Biases

*One of the few efforts under way in the schools and in public discourse to develop stage 4 literacy—the ability to see, evaluate, and control tech-nology—involves media literacy.*

*As the term suggests, the effort concerns information, news, and entertainment media rather than all of technology. This is an obvious place to start, given that our living rooms and class-rooms are awash in media.*

John Davies's media myth no. 1: The media tell us how life really is.

*Advocates of media literacy are split between those who focus on the media and those who focus on media's content. The content-oriented advocates seek more con-trol over programming, while the media-oriented advocates seek to raise the public's awareness of inherent media biases. It is the latter effort, the drive toward develop-ing the skills necessary to understand how the inherent biases of media distort the images and messages they convey, that concerns us as STA investigators.*

*A medium may manipulate its messages in at least four ways that invite our examination: editorial, personal and cultural, technical, and commercial.*

Editorial. *Print publications, radio and television stations, and other media present the world through their own ideological lenses, with or without malice aforethought. They may be conservative, liberal, or nonpartisan; anti-abortion or pro-choice; religious or secular; pro-development or pro-environment; and so on. Some media organizations are up-front and proud of their biases. Others feign to be objective but are not. The presence of such slanting makes it unlikely that a medium will have the kind of balanced perspective sought by advocates of media literacy.*

*Davies's media myth no. 2: The media keep us well informed.*

Personal and cultural. *Whether or not those who create and interpret media have official biases, they have personal biases rooted in the culture in which they were raised or now live. Despite their best efforts they are seldom able to convey renderings of reality independent of their personal biases. Furthermore, as Margaret Wheatley reminds us so eloquently in her book* Leadership and the New Science, *when it comes to human events—even in the sphere of business and commerce—there is no objective reality independent of the relationships that define it and the people who observe and interpret it. It is always fair to ask which personal and cultural biases have crept into a media presentation and the possible extent of their impact.*

Technical. *Every medium favors particular kinds of content, a quality often referred to as technical bias. Here's one example. Because television favors the transmission of movement and confrontation in order to hold viewer interest, reporting is slanted in that direction. Thus a day-long peaceful protest may be presented on a primetime news show in a fifteen-second clip showing the only scuffle that occurred. No matter how much it distorts the event it is used because it makes good TV— that is, it follows the inherent technical biases of the medium, biases that have gained more importance in TV journalism than the objective portrayal of truth.*

Commercial. *Media presentations are bought and paid for largely by advertisers. Therefore commercial sponsorship to a great extent determines, either directly or indirectly, what the media convey. If no corporation or patron wants to sponsor a TV program that takes a hard look at agribusiness, the program's viewpoint probably won't find its way into the public discourse. Anyone seeking to instill awareness of media literacy must keep this situation in mind.*

*Davies's media myth no. 4: The media provide a free flow of information.*

## + EVALUATE

We ask:

- What behaviors does the technology promote or prevent? Discourage or encourage?

- How does the technology change our capabilities and predispose us to certain actions?

- What does the technology cause us to do that we weren't doing before we adopted it?

- What social conditions are necessary to create and sustain a particular technology?

- What content does the technology favor?

**How is the technology interpreted?**

### SEE

So when is the message the message and not the medium?

*What you get is what you see.*

The hammer in my hand may make me a hammerperson, but whether I use the hammer as an instrument to build a house or a weapon to do someone serious injury depends on how I interpret and apply my hammerness. As the saying goes, a knife can cut both ways.

A STOP sign bolted to a metal post at a street corner means "stop" to some drivers, "slow down" to others, "speed up and get through the intersection" to still others. A STOP sign bolted to a wall in a fraternity house represents a trophy to some, a deconstructionist piece of kitsch decor to others, theft of public property to the morally minded. To the fraternity brethren, it doesn't even imply stop or go.

As I attempted to show in Chapter One, if we look closely at our machines and our artifacts we can see our own reflections. They are trying to tell us who we are by virtue of how we interpret them. As we use, adapt, stretch, and recast our artifacts into behavioral statements, we broadcast who we are through the choices we make about them.

This mirror-like image contrasts sharply with the primal, unconscious world of hidden agendas that dominate the domain of implicit technological biases and technological politics. In that murky, subterranean world we rarely *choose* anything because we rarely *see* anything.

It is not the case, of course, that all of our technological interpretation is conscious—far from it. Habit and conditioning render much of what we do invisible to ourselves unless we care to examine our actions. The point is that if we have the presence of mind to look, we can explore our technology all we like—and, in its mirror, get a better look at ourselves.

That gives us a fighting chance to pull the relationships we have with our artifacts within the purview of stage 4 literacy, where we can see, evaluate, and, let us hope, control them. As STA sleuths we need to be able to distinguish human interpretation of a technology from implicit behavioral bias.

### + EVALUATE

We ask:

- How do we interpret the technology?
- Do other people interpret it as we do, or is our interpretation highly personal?

### Who is selling the technology and why?

### SEE

Not all "selling" of technology is done for purely monetary reasons. What other reasons may there be? What about promoting the public interest? What about promoting a cleaner environment? What about selling spiritual salvation?

I still pause when I find a Gideon's Bible in a hotel room, one of the few true giveaways in the world. I find equally impressive the Bible-bearing proselytizers who appear at my door and never ask for money, obviously motivated to share their technology—the printed page—by something other than financial gain: devotion to a belief.

But the fact is that very little technology appears in our lives without a push from the marketplace, which spills over with too many goods chasing too few motivated customers. While endless innovation and overpopulation might seem to mesh nicely, the reality is that each human being has only so much shelf space at home and limited attention to devote to researching how to fill it. Those whose job it is to convince us to buy things therefore must speak deftly and quickly lest we change the channel.

### + EVALUATE

We ask:

- Who is selling the technology and why?

- Are the reasons purely monetary, or do we detect devotion to an ideal, cause, or perspective in the promotion of the technology?

### How is the technology sold?

### SEE

Advertising is such a sophisticated, complex phenomenon that a fair treatment of it would fill more pages than I can devote to it here. Still, certain self-evident truths about advertising sales-

manship can help us, as STA investigators, lift the curtain and
see the wizard behind the facade of Oz.

*Advertising is a question of how, not why. The why is money. The how is craft.*

First, advertisers, like lawyers, aren't paid to seek the truth;
they are paid to convince others to adopt a particular point of
view. To accomplish this feat advertisers expertly use what we
often accept as the tools of truth, that is, media, along with
media's most talented manipulators—actors, athletes, com-
mentators, doctors, psychologists, and other salespeople—to
convince us of their truthfulness.

Thus we need to judge the content of a commercial or
advertisement against media's power of convincibility. What
some smooth talker passes on as truth may appear so only
because of its enticing presentation. The stage 4 literate always
needs to be on high alert—seeing, evaluating, controlling.

## Hold the Waffling

*Let's dwell for a moment on our culture's devotion to presentation at the expense of content.*

*It is at the heart of our adversarial legal system, which often trumpets client satis-
faction and professional triumph over truth and justice.*

*It is also becoming endemic in our education systems. In language arts, for exam-
ple, we often fail to encourage students to report on both sides of an issue but ask them
to "develop an argument" to convince others of a particular point of view. Indeed, presenting a balanced argument is often seen as "waffling."*

*The situation has become scary because of the persuasive power of the information
technology that is beginning to permeate our schools. It was one thing to present an
argument in a handwritten text, emulating legalitarians in their quest for enlightened
one-sidedness. But as the term paper yields to the multimedia presentation, in which
students present their positions with the full force of audio, video, music, graphics, photography, and animation as well as text behind them, "developing an argument" begins
to look suspiciously like advertising.*

*The philosophical undercurrent at work here is "convince me" when it should be
"inform me." This undercurrent has been with us as a cultural propensity for a long
time, but postmodern technology greatly amplifies it.*

The second self-evident truth is the first rule of salesmanship: don't sell people what they want; sell them what you have.

*Caveat emptor:* ours is a world in which the buyer is warned to beware. It is a world committed first to making the sale, secondarily to other issues, such as safety, good workmanship, ecological integrity, cultural displacement, and personal alienation.

While adults may reasonably be expected to beware, children rarely have the tools of discrimination necessary to do so. Simple: buyers can beware for their offspring as well as themselves. Unrealistic: children bathe in media from sun up to sun down. Filtering data for yourself is a full-time job. Doing it for someone else will produce mediocre results at best.

The third self-evident truth is that advertisers rarely sell the thing itself. Instead they sell positive or desirable feelings and impressions that they seek to attach to the product. Enamored of the feelings, so the theory goes, customers will buy the product by association.

Conventional Madison Avenue wisdom says that while we might be reluctant to buy cars, cigarettes, or mechanical toothbrushes on their own merits as quality products that serve compelling needs, we will more or less unconsciously buy the sex, pleasure, and status offered by these products when they are presented with the right glitz.

Beyond the purely psychological considerations of selling lie the mechanics. This subject has become particularly interesting given the capabilities of satellite and cable TV, Internet databases, and ecommerce Web sites that ooze gobs of data on consumer spending preferences.

Scattershot once was the preferred methodology of advertisers. They aired a commercial for Gillette during the Super Bowl hoping it would reach a sizable market share even though bearded men, children, women, and other nonshavers might also be watching. Postmodern communication technology has created two new methodologies that significantly decrease the salesman's guesswork: niche marketing and pointcasting.

Niche marketing responds to detailed market analyses of audiences and makes sure that only those who can afford Mercedes Benzes receive ads for them. Everyone else belongs to Ford country. If I fill out one of the many tracking devices that marketers disguise as warranty cards or user surveys, in which I reveal my job, salary, and purchasing predilections, I can expect focused junk mail to follow.

*Davies's media myth no. 8: Consumption is inherently good.*

Pointcasting takes narrowcasting a step further, using the Net to establish a one-to-one relationship between myself and an advertiser. Most search engines, as they are known in Net parlance—sites such as Yahoo! and Hotbot that find information in the overgrown data thickets of the Net—also feature advertising windows whose content changes based on the subject of my search. As they know more about me, they send—or push—information to me in the form of email, mailed catalogs, even automated voice mail.

In essence there is nothing wrong with that, particularly if I need what's being pushed. I am free to ignore all the advertising and go straight to the information if that's what suits me. But I await the day when information merchants reverse the business dynamic that assumes they have to hide from me in order to be effective.

I should be able to hire them directly as my information agents, expecting them to keep my résumé, buying interests, and credit history in a safe place and give them out only to those who are selling what I'm interested in purchasing. They could just beep me now and again and ask, "Since you just bought a new house, would you like to be put on the mailing lists of landscapers?" They could inform me that a company wants to add me to its database, then let me decide whether I'm interested. And as my interests evolve I should be able to tell them so they can redefine the ways in which they filter the avalanche of information sent to me.

*Taking people's cars out for a spin without asking is a crime; taking their information is not.*

But that is far from the way it happens. Readers in the mood to boost their suspicion of the entire information underbelly of our free society might want to check out Jeffrey Rothfeder's *Privacy for Sale: How Computerization Has Made Everyone's Life an Open Secret*. The fact is that in *this* free society, my atoms may be mine, but my bits aren't.

Bottom line: Advertising is acting. It is innuendo. It is strategy. It is multimedia-driven and multidimensional. It is powerful. And, given the nature of the new technologies, it is personal.

## Search and Ye Shall Find

*Not long ago I was searching the Internet for information on gardening. Within seconds, along with a multitude of hyperlink references to Net-based gardening information, I was also looking at a dancing ad for garden supplies, books, and other such consumables.*

*I get it! The advertiser is having a conversation with me, assuming there's a commercial angle to anything I want to know.*

*That is the philosophical foundation of push technology: everything can be commodified. And promoters, rather than waiting for me to ask for something, see to it that it is sent to me automatically. Sort of a consumer guardian angel watching over me to make sure I don't miss a buying opportunity. Yahoo!*

*Dare I do a search on Mother Teresa?*

### + EVALUATE

We ask:

- How does the product's promoter find potential buyers?

- Which needs does the product's promoter claim to fulfill and which are actually fulfilled?

- What images does the product's seller project about it and what techniques are used?

- In what ways does the advertising sacrifice truth to image and presentation?

- What is the true objective of the images presented?

- What is the advertiser *not* telling us?

**What is being commodified?**

**SEE**

We want feelings, but we can't buy feelings. We can only buy things. So it is up to product developers and advertisers to package feelings as things in order to create a common currency shared by producer and consumer.

Creating such currency constitutes the crux of the advertiser's craft, a process known as commodification. To pierce the process that leads to the presence of technology in our lives, we must understand how feelings masquerade as things and why they wear the costumes they do.

We don't buy the car in the commercial based on an objective rendering of its value. Such a rendering would be impossible to provide in a thirty-second spot anyway, even if we watched it with rapt attention. Instead we buy the beautiful, smiling people in and around the car, the party it escorts us to, the ticket to high society it implies, the lost sensuality it will restore, the liberation it guarantees — all of which are commodified as a car.

But we are still dancing on the surface. Much of this is metaphor. Commodification also happens on a much deeper level.

As we delve deeper it becomes apparent why advertising attracts artists. Art and advertising both consider themselves mirrors — lies through which we see the truth, to paraphrase Picasso. The difference is in their aims. Art seeks to disturb; advertising seeks to please. Art seeks to show us as we are, advertising as who we want to be. Art wants to make our psyche bleed; advertising wants to cauterize it, massage it, console it.

Art and advertising do, however, share a similar understanding of the human condition. When the vigilance of the neocortex wanes and stage 3 literacy beckons us to forget the present and visit the evolutionary past that lives on only in our brain, the advertiser and the artist alike see a mix of what Stephen Fox, in his book *The Mirror Makers: A History of American Advertising and Its Creators,* calls "materialism, sexual insecurity, jealousy, vanity, and greed."

*The difference between pain and pleasure is largely a matter of salesmanship.*

Those traits are uncomfortably close to the seven deadly sins: pride, envy, gluttony, lust, anger, covetousness, and sloth. They are the aspects of the human condition that the artist seeks to expose. And they are the raw materials that the advertiser seeks to commodify.

It's not pretty. And it's important to remind ourselves that the flaws which attract advertisers are not the sum total of who we are, merely our most exploitable weaknesses. We seek to understand this part of our psyche not to make ourselves feel any worse than we already do but because advertisers understand it so well. They attack our unconscious, flying beneath our stage 4 literacy radar. As conscientious STA investigators, we must do whatever we can to be ready for them.

## + EVALUATE

We ask:

- What human desire or weakness is being commodified?
- What form does the commodification take?

**Is the need for the technology real or created?**

### SEE

The last stop on our tour of the human psyche as seen from an advertiser's perspective focuses on what artists call tension resolution and product managers call need creation and fulfillment. At the heart of this issue is whether the need that a technology seeks to satisfy is real — that is, generated naturally by the consumer without a push from the marketplace — or created by the advertiser and product packager.

*Much of what passes for choice, Langdon Winner concludes in* Autonomous Technology, *is just adaptive response.*

One view holds that an identifiable need already exists prior to the creation of a technological solution and that product developers are, once again, simply mirroring the human condition and attempting to satisfy its wants. Henry Petroski, in *The Evolution of Useful Things,* makes a compelling case for the proposition that most artifacts have met human needs. We have four-pronged forks because less-endowed forks failed to satisfy our quest to eat effectively and in a dignified manner.

But while forks are one thing, Beanie Babies are another. Jerry Mander contends in *Four Arguments for the Elimination of Television* that most perceived needs for the goods we buy are artificial. The advertiser's art, he admits after years in the advertising business, does not lie in making us buy something we don't want but in selling us something we don't need.

As Winner suggests in *Autonomous Technology,* "with each new invention it becomes possible to awaken and satisfy an appetite latent in the human constitution." And, he adds, apparently "the human being is by nature a creature of infinite appetite."

### *Did I Smell Something?*

*Advertising agencies generate our needs in order to sell us a solution to a problem we didn't know we had. That might be called reverse engineering, working backward from the product to the desire.*

*A car satisfies obvious needs, right? After all, what are roads for?*

*But a heavily scented deodorant may be a solution to a problem that didn't exist until advertisers created it.*

*Look around you. How necessary is that fourth clock radio you have? How much would you miss it if it disappeared?*

*This kind of introspection is not intended to make you feel embarrassed about your level of consumerism. It is meant to spur your thinking about where the need for what you own came from.*

*It is possible that we need everything we have. It is also possible that we have been convinced to need what we have simply to keep the wheels of industry rolling.*

Winner explains that a need becomes a need in substantial part because a megatechnical system external to the person needs that need to be needed.

## + EVALUATE

We ask:

- Does the technology meet real human needs, or does it create new needs?
- Who defined or tailored the need? How?
- How well did we function without the technology?

## The Effects of the Technology

Our technological beast shakes off the lingering effects of the tranquilizer, looks about in dazed distrust, and lumbers back into the world of human events.

As it does so, waves of change ripple outward, smashing against some parts of life with the force of a tsunami, washing over others imperceptibly like calm seas nipping away at the shoreline. Some of its impacts are delayed, awaiting a long chain of events to unfold before we can understand what we wish we would have understood sooner. Others will happen far away, as a distant culture begins to slowly adapt to the philosophies implicit in someone else's technology.

The beast affects every part of life, because there is no such thing as "the technology itself."

In truth, as we consider the effects of technology we have not abandoned the reductionist perspective entirely; we have merely cut the world into larger slices. We focus on technology's impacts from three general perspectives: the environment, ourselves, and our institutions.

Anthropologist Leslie White, in *The Science of Culture: A Study of Man and Civilization,* describes culture as consisting of three strata, with technology on the bottom, philosophy on the top, and we the people in the middle. In White's mind, we represent a compromise between what we can do with our tools and what we aspire to become as a result of our mythology. Here we focus our questions on how we, the middle layer, play out our hand in relation to all three strata.

## How does the technology affect the physical environment?

### SEE

We could bring on nuclear winter, unleash legions of lethal liquids into our lakes, terrorize our terrains with toxic trash — and the environment would get along just fine.

Of course we'd all die, and so would much of what is familiar and valuable to us. But Mother Nature, with billions of years of sunshine left and eons of rock-steady patience on her side, would mutate and evolve and eventually produce a

*As Neil Postman observes, a new technology doesn't merely add something; it changes everything.*

*Oceanographer Clay Good warns that many other life forms stand in line to fill our niche should we make it uninhabitable for ourselves.*

world we would call — were we alive to witness it — fascinating, robust, natural, even sacred.

Most people concerned about the environment are really concerned about themselves and other human beings — as they should be. It is an honorable and pragmatic perspective to see the destruction of the environment as slow suicide, though not nearly as humble, eloquent, or informed as Native America's warning to European settlers that the land owns them, not the other way around, and that their failure to see things from this perspective was symptomatic of the vanity that undermined their chances of survival. But it is perhaps the best tack that technological humans can take for now.

*There's no such thing as stepping lightly on the earth.*

How do we discern the impact of a technology on the environment? We need to consider its entire life cycle, which includes making it, selling it, buying it, sending it to the people who use it, then throwing it away.

To begin this exercise in stage 4 literacy, we zero in on phase one of this process, which we will call production, by working backward from a microwave oven, for example. As STA investigators, we take an imaginary trip through its components and subsystems to its ancestry: the sand for the little microchips, the various metals used for its casing and internal components, the plastic for the handle, the paper on which the instructions are written, the ink that forms the words on the page. What are their origins?

We see the microwave oven as the end result of resource extraction (harvesting, mining, drilling, chopping, shoveling) and resource reshaping (smelting, pressing, molding, chemical engineering, manufacturing), both of which represent substantial interactions with the physical environment. Developing stage 4 literacy means replacing the magic of the finished product with knowledge of its origins.

## Give Us This Day

*A friend of a friend once visited me in my rural home, a young man who was making his first escape from the city for an extended period. He admitted to me that his life consisted of little more than his home in suburban New Jersey, his office on Wall Street, and the mass transit system that connected the two.*

*As we sat discussing the world at a picnic table in my front yard, he gazed across the road at a field of wheat, its stalks wafting gently in the afternoon breeze, and asked me what it was. "Wheat," I told him.*

*After a long pause, his face contorted and his voice rising in disbelief, he asked, "As in bread?" When I assured him, "Yes, as in bread," I could almost see the magic of the presliced, cellophane-wrapped, roast beef and mustard container leave his conscious-ness. Wheat ... bread. The sorcery was gone.*

How do we end such magic in real terms? It helps to take a look at the big picture. As a teacher I have come to appreciate the value of getting metaphorical. Toward this end my students and I spread out a big piece of paper—big paper, big picture—usually newsprint, which one can buy cheaply in stores or obtain free from a newspaper publisher. Before continuing we pause for a moment to acknowledge the resources composing the paper that have become sacrificial lambs in our quest for understanding.

*Any sufficiently advanced technology is indistinguishable from magic.*

—Arthur C. Clark,
*How the World Was One*

Then we get kinesthetic. I have also come to understand the value of getting physical—or "appealing to the kinesthetic mode," in pedagogical parlance. The body often remembers what the mind forgets. My students and I do it by spreading the paper out on a table and beginning a process in which we draw on it, walk around it, sit and stare at it, write on it.

*We need plants, but they don't need us. We have the same relationship with technology.*

First we draw a small circle in the center the size of the bottom of a coffee cup. In it we write the words "Microwave Oven." Then we draw and label a series of concentric circles that represent the oven's life cycle, increasing the distance between circles as we draw them: "Production," "Distribution," "Selling," "Consumption," "Waste."

Then, using the ability to see and evaluate that we developed in considering the impacts and goals of the technology as outlined in the first two sections of this chapter, we look for the effects of everything—from the plastic-wrapped heat-and-serve lasagna that often inhabits microwave ovens to the newspaper advertisements and TV commercials for both the microwave and the lasagna to the electricity needed to run the microwave, the newspaper presses, and the television. The ripple effects move ever outward, forming a tecosystem that seems as extensive, intricate, and elusive as nature itself—a tecosystem viewed from the perspective of impacts and effects.

We investigate the ways in which all of these things extend us, how they impact our temporal and spatial appreciation of life, how dependable they are, and all the other characteristics of "the technology itself." We write down all our findings within their appropriate circles, filling our big piece of newsprint in no time.

*Ivan Amato estimates that humans extract fifteen billion tons of raw material from the earth each year.*

When we have finished writing, we multiply the information on the large piece of paper by half the number of friends we have to get an idea of the amount of natural resources that are devoted to microwave ovens within our own community.

---

### Going to the Sources

*You too can zoom out and see the big picture of a technology.*

*You will find a number of sources of information: consumer groups, the Environmental Protection Agency, countless books and magazine articles, even the technology's manufacturer, who may offer good, though often contradictory, information.*

*The Web is a good place to look if you know what you're looking for, but beware of the land mines of unjuried information. Two adages come to mind: "Consider the source" and "Don't believe everything you read."*

*Above all, as you get scientific you should straddle the line between needing proof and trusting your intuition. Both are valuable.*

## + EVALUATE

We ask:

- What are the documented effects associated with the life cycle of this technology, including the effects of using the energy source it requires?
- What is our intuitive understanding of these effects?
- Do the documented effects and our intuitive understanding agree or clash? .

### How does the technology affect the human body?

### SEE

The most relevant and powerful effects of a technology usually are local, and there is nothing more intensely local than oneself.

*Our health seems much more valuable after we lose it.*

*—Author unknown*

Much of the process involved in looking at the technology's effects on the physical environment also applies here. We see our own body as a subset of the physical environment, one more tree drenched by the forces of technology as they wash over the jungle.

Each circle that emanates from the microwave oven produces a localized effect on me and you, the users — to borrow a word that has gained notoriety from the politics of addiction. While many effects are obvious, some are subtle enough to warrant special attention if we, as STA investigators, are to pull them into the light. Detecting them requires stage 4 literacy.

Edward Tenner calls them revenge effects. In *Why Things Bite Back: Technology and the Revenge of Unintended Consequences,* he tracks technology's hidden agenda, a sort of Newton's third law of motion for machines and technological systems. For every intended consequence of technological development, he argues, an unintended revenge effect occurs naturally to offset it.

One of technology's favorite ways of doing this, according to Tenner, is "to replace life-threatening problems with slower-acting and more persistent problems." Antibiotics, for instance, clobber a short-term infection but leave the victim more vulnerable to disease in the long term. That apparently is nature's way of reminding us there's no free lunch.

Tenner says the same is true of office technology, which "had promised to make work painless [but] unexpectedly attacks muscles, tendons, and vertebrae," producing the maladies of the Information Age worker. Tenner finds equally ironic examples of unintended consequences in the use of protective athletic equipment, in transportation advances, and in other areas of human endeavor.

Nonetheless, in all of it he finds cause for cautious optimism. Through disaster we generate awareness, which leads to renewed vigilance, which begets the engineering of precaution—a kind of stage 4 literacy for developers and inventors.

We can never escape revenge effects, but we can seek to contain them as much as possible by simply paying attention to what we make and how we use it. We can never gain perfect control over our technology, but we can use every downside to bring us closer to an upside.

Yet there looms the mother of all revenge effects, an ultimate reckoning one day that makes stage 4 literacy a painful endeavor: overpopulation, along with the resource depletion and environmental degradation it spawns, brought about primarily by a tecosystem committed to making life longer and more comfortable. But what can one say against artificial limbs, pacemakers, and contact lenses?

Recalling the opposing viewpoints of the philosopher and philosophee as listed in Chapter One, we force ourselves to trudge up the hill to the ledge of objectivity where we can see the growth in population, elongation of life, and rising standard of living that result from these technologies. We look out over the masses of people with their vastly unnatural yet vastly improved lives, see the inevitable clash between individual rights and availability of resources, and worry. Some day, it seems, the bills will come due, and we, intent on our quest to evolve an egalitarian and humanitarian society, may be unable to pay them.

## More Sources

*As with the natural environment, a great deal of documented research exists concerning the effects of technology on the body.*

*Physicians, books, journals, health organizations both orthodox and alternative— all have valuable information to impart, often without charge.*

*Here too the Web amazes in its ability to offer access to highly focused, though often unjuried, information. As always, seek the truth but feel free to assess issues in the light of your own experience and intuition.*

### + EVALUATE

We ask:

- How does the technology improve the lives of humans and the life of the environment?
- What revenge effects await us?
- What are the long-term costs of life-extending technologies?

### How does the technology affect social relationships?

### SEE

Much has been written about the immense power of technology to rearrange social relationships. In *No Sense of Place,*

Joshua Meyrowitz states that we can consider media as "certain types of social environments that include or exclude, unite or divide people in particular ways." It is no different for other technologies.

We can look at any technology as the third party in a *ménage à trois* — us, our personal community, and our gadgets. Anything we introduce into our social life — a pet, a computer, a car — can't help but affect our interactions with others, our priorities, our loyalties. It is a question of how, not whether.

To understand this truth, we must look at the changes in our lives. Technology is advancing fast enough that we can serve as our own before-and-after gauge.

Focus your attention on that new gadget. First think back — what was life like pre-gadget? Then think ahead — will you still be using that gadget next month?

Be aware. Think about the changes. File them. Revisit them now and again, asking yourself whether you miss the experiences you had pre-gadget or prefer the experiences that replaced them.

*According to Sherry Turkle, we use our relationships with technology to reflect on the human. We seek out the subjective in the computer.*

## Going Espresso

*Every gadget connects and disconnects, ultimately recontextualizing us socially.*

*Before you fire up your new espresso machine — that is, before the gadget recedes into the background of your life where you can no longer detect its presence, let alone its impact — consider how it will reduce those pleasant interactions with your acquaintances at the coffee shop.*

*Or, on the plus side, consider whether you might be cultivating new relationships — say, with espresso aficionados, coffee suppliers, or those online acquaintances with whom you interactively chat as you sip your home-brewed café au lait.*

*Consider not only what the espresso machine costs you in dollars or offers you in convenience but also what it might cost or benefit you in social connections.*

The analysis of technological impacts on social relationships often concerns the family because we look to the institution of the family to provide bearings in a world that would otherwise be social chaos. Family serves as a container, and when it cannot contain the events and emotions that have traditionally been its domain, they leak out into our neighborhood, community, or legal system.

The assumption about technological interference often seems to be that when people interact less, life is diminished. It is the subtext in McLuhan's de-romanticization of the book as technology, which he saw as forcing people to forgo storytelling and group interaction in favor of a solitary activity and a private point of view.

I find this purely quantitative approach academic at best, useless at worst. We all know many solitary activities — painting, jogging, meditating, playing the piano — that are healthy.

## *TV or Not TV*

*Marie Winn discovered a simple way to bring peace to the family.*

*In* The Plug-In Drug: Television, Children, and the Family, *Winn tells what happened to families who, as an experiment, turned their homes into television-free zones. They discovered a greater closeness among family members, a more tranquil atmosphere in the home, more help from the children with the housework, and greater engagement in cooperative activities — with no apparent downside at all.*

*Yet, Winn reports, most of the families eventually returned to TV land. Why did they leave paradise once they had found it? Winn postulates a "need for passivity, for self-annihilation, for regression to a state of dependence." She adds: "Perhaps a life of activity, of self-searching and growth, seems too difficult to attempt in our fragmented society."*

*The withdrawal of a technology no doubt amplifies the human condition, which, ever in a state of flux, cannot maintain peace indefinitely. Or perhaps the families who turned their TVs back on simply missed their fellow global villagers.*

And we know many unhealthy social situations that we would rather avoid. Kurt Vonnegut calls them granfalloons—gatherings, such as family reunions, of people who are compelled to be together but would rather not be (see Vonnegut's collection of essays, *Wampeters, Foma, & Granfalloons*).

Vonnegut's cynicism about family reunions notwithstanding, I find his point well taken. When it comes to social relationships, quality not quantity is the issue. More time together can mean more chances for war as well as for peace.

*Program concept for Jerry Springer: The plight of children who try to love cross-platform parents—Mom's got a Mac, Dad's got a PC, and they just don't get along.*

### + EVALUATE

We ask:

- Does the technology make us more involved with the people around us—or less? How does it do this?

- Does the technology improve or diminish the quality of our interactions with others? How?

- Does the technology promote or preclude group activities? Which kinds?

- Does the technology alter core family relationships? How?

**How does the technology affect work?**

### SEE

If I stare hard into my cup of coffee I can see underpaid workers toiling in substandard conditions picking the beans from which the magic liquid is made that helps bring me to life every morning. So I avert my eyes.

*Through a willing suspension of suspicion, the drama of technology plays out.*

Behind every "thing" there lies the story of how it came to be—the story of work. One of its main plots is the relationship among the three primary players of any economic drama:

labor, management, and consumers. We are concerned here with how technology restructures, amplifies, and often distorts these relationships on two fronts: industrial and digital.

Despite the growing preponderance of bits in the workplace, we will always need an "atom-based" industrial sector to manufacture computers and modems that supposedly brought the Industrial Age to an end, as well as such commonplace accouterments as clothes, cutlery, and cars that ground us in the past. What differs now is the rate of change and the intensity of work-shift as the pace of technological evolution accelerates.

## Robotics @ Work

*Consider the impact of robots on industrial manufacturing.*

*Robots replace the assembly-line worker, but they provide work for robot designers, producers, operators, and maintenance mechanics.*

*Management, enamored of the structural beauty of efficiency and looking for bottom-line vindication, embraces robots because of their impact on price, reliability, and quality, regardless of the restructuring of the drama of work. Workers, by contrast, differentiate the role of the human with dignity and a family to feed from that of a cog in a machine.*

*Push, pull.*

*We mediate the two viewpoints by continually redefining our roles through the lifelong learning process, a continual rescripting of the drama of the workplace.*

*But as consumers we go about our shopping fairly oblivious to it all.*

If we press our ear to the wall we can hear the din of tension between workers and management pursuing different, often mutually exclusive goals. Whenever we hear the economic deck being reshuffled we should look about to see who is winning and who is losing.

The employment of robots is a case of using new technology to do the work of older technologies. But going virtual provides the flip side, at least philosophically, of a fragmented industrial drama: convergence. Here we truly have a new storyline with a new plot and new protagonists.

The bit-based knowledge industry, requiring no heavy lifting, does more than the hydraulic lift to equalize labor among the ironman and the hundred-pound weakling. Unlike the world of the lift, idea work favors the clever. And where brain power overtakes brute power, the living room with a laptop computer becomes the locus of innovation.

At the root of the changes are the new relationships brought about by a flow of information and a multiplication

---

### The Wordminters

*In* Digital Economy: Promise and Peril in the Age of Networked Intelligence, *Don Tapscott describes an ethereal workplace that exists globally and immediately on the Net, creating relationships so new that we have had to coin a new vocabulary to describe the results.*

Prosumption *involves buyers in the design of their own products to ensure that they get exactly what they want. Acting prosumptively, consumers don't need to settle for whatever appeals to mass-market tastes. They can have any movie they desire on demand, clothes to fit their unique sizes and shapes, or custom-designed kitchens just by using arcade-like computer interfaces.*

Disintermediation *uses the Net to eliminate the maze of product handlers, the "agents, wholesalers, distributors, retailers, brokers, or middle managers" who massage products on their way to the consumer. Disintermediately, consumers buy whatever books they want, purchase the least expensive airline tickets, hire people based on Web page résumés, renew their driver's licenses.*

Adhocracy *is the term used by Robert H. Waterman in his book of the same name to describe the system that replaces the hierarchy as the organizing principle of the new corporation. The adhocracy aligns the knowledge resources within an organization on an ad hoc basis to solve specific problems, bypassing the messy one-size-fits-all chain of command that restricts the flow of talent and ideas.*

---

of information sources that didn't exist even ten years ago. Still, the middle managers are not dead; they are just being reborn in a new guise.

Donald Norman laments that "a place for everything and everything in its place" works only if you can keep track of the places.

The new middle ground is being populated by people who make sense of the Net for those of us who don't have time to do so. Between us and our information will stand more and more of the agents we need to understand the information that, if we're lucky, will liberate us.

The digital domain of work sounds great—and it is if you have the means and the opportunity to participate in it. Yet if I stare hard into my computer screen I can see more underpaid workers—some of them halfway across the world, more and more of them closer to home—toiling in substandard conditions for subhuman wages assembling the technology that does the digital magic of delivering my mail each morning as I sip my coffee.

### + EVALUATE

We ask:

- Does the technology weaken or strengthen the power or bargaining position of management? Of workers? Of consumers?
- How does it impact working conditions?
- How does it raise or lower people's standard of living?
- How does it change the nature of the interactions among producer and consumer?

### How does the technology affect self-relationships?

### SEE

Here we are concerned with spending time alone understanding ourselves. As serial-processing human beings capable of

Educational technologist Michael Byer notes wistfully that technology inhales time.

attending to only one thing at a time in any great depth, we examine the consumption of time for which technology is so often noted.

Technology either helps us find our way or gets in our way.

The buzzword among technologists for technology's pathfinding prowess is *empowerment* — the use of technology to leverage our abilities and dreams. Good examples are the multimedia tools that assist the artistically challenged. Such tools can open doors for many persons unaware of the passageways to self-expression, marginalizing the need for manual dexterity and providing editable canvases that allow almost anyone to express a personal artistic vision.

When technology only gets in the way we become "distracted from distraction by distraction," like the shallow modern human of the earlyTwentieth Century described byT.S. Eliot in "Burnt Norton." Distraction is not new; people have been avoiding themselves for millennia. Technology simply facilitates it to such a high degree that every now and again we need to check our interior landscape, just as we need to check our exterior landscape, to see what has become background.

Out of control and too busy to notice.

Because we can do only one thing well at a time we are forced to prioritize, make decisions, and define ourselves as much by what we don't do as by what we do do, so to speak. Life within this domain is often just a sophisticated time chase within a closed system in which we use technology to liberate minutes that are almost immediately filled with other technology-dominated activities.

By default the goal within this system often becomes engaging in beneficial versus wasteful activities. The microwave oven that allows me to cook dinner faster and spend less time connected to the rites of sustenance liberates minutes I can spend at my computer in creative self-expression or playing games.

They become my own hard-won minutes, and how I spend them is entirely up to me. One person's computer game is another person's creative enterprise.

But it is important to understand that the decision making often happens within a technologically mediated culture of excitement. Such a culture makes it difficult to see that the moments spent within it could also be spent outside it.

It is even more crucial to understand that the exterior of the tecosystem is often the interior of one's consciousness.

*Filling ourselves vs. fulfilling ourselves.*

Leaving the bubble doesn't necessarily mean taking a walk in the woods. It quite often means traveling the privacy of one's inner landscape — praying, meditating, looking for clues about who I am, what I am missing, and what has become of my dreams. It also quite often means examining what has become of one's mythology and who have become one's gods, heroes, and demons.

Mary Ann Lieser turned off the technology in her life. "With nothing to fill the silence but my own voice, I began singing," she writes in her essay "The Media Free Family," in Bill Henderson's *Minutes of the Lead Pencil Club: Pulling the Plug on the Electronic Revolution*. As STA investigators, we want to know whether our technology keeps us from singing or helps us to find our voice.

## +EVALUATE

We ask:

- How does the technology change who we are, how we see ourselves, what we expect of ourselves, what we are capable of doing?

- If the technology gives us more free time, how do we spend this time?

- Does the technology help make us more or less expressive?

- Does it put us more in touch with ourselves—or less?

- Does it keep us from experiencing gods and heroes, or has it become our god?

## How does the technology affect the power structure?

### SEE

Given that technology offers the tools of upward mobility, the issues of empowerment and control are inextricably interwoven and deserve special illumination. We are concerned here with how people use technology to exert influence over others, as well as how it thwarts their attempts to do so.

Every technology we have extends our powers beyond the reach of those who don't have it. While the same technology may also disconnect us in many of the ways already discussed, we focus almost entirely on the fact that the technology provides the "reach" we want our children to have in order to grab the opportunities we hope they want.

In the land of Nod—the forgetful world of stage 3 literacy in which the amount of technology that has accumulated in our background defines our place within society and the only real crime is failure—reach is the standard of measurement we use to determine our place in the power structure. An honest examination of the technology I have in my life may reveal that I have more power than I want or need. And it may also reveal that someone close by needs the power that I have and often don't use.

Exploring the concept of power forces us to come face to face with one of those nasty, ancient, perhaps genetically predisposed limitations of the human spirit. I'm speaking of our tendency to abuse power given the chance, a trait mightily amplified by technology.

We can see this trait vividly expressed if we examine the interactions of a country's information media and its government. The tendency toward abuse is bad enough in a more or less open democracy (if anyone had doubts about that, Monicagate surely dispelled them).

But abuse flourishes particularly well in a closed system that excludes the masses and discourages them from communicating among themselves.

When I visited the Soviet Union in 1988, I was surprised to find that most of my colleagues in Moscow didn't have phone books. Someone described the situation to me as an ingenious way for the government, using the pretext of conserving paper, to keep people from organizing.

*As the distance between the powerful and the powerless grows, Michael Goldhaber warns, recognition of their common humanity becomes more difficult.*

In the United States we usually operate under the assumption that the more information the better—and to hell with the trees. The information inundation we demand sometimes acts as a floodlight, illuminating everything all at once and making it impossible to hide anything, whether or not anyone is looking for it. Where the floodlight is aimed depends, of course, on who controls it.

The power of mass media to favor the very few over the very many, combined with the inherently insecure, power-abusing nature of *Homo sapiens,* leads democracies to feel more comfortable with distributed rather than concentrated power—at least in theory. That is why, after four decades of the monocultural, centralized authority of broadcast television, we applauded the coming of the Net, which began as a Defense Department experiment and leaked into the public domain as an uncontrollable contagion of free association and unfettered speech.

Most people understand communication technology's potential to strengthen democracies. It can pierce layers of bureaucracy with a single email message. And because Information Age workers remain unseen, we can be stripped of our narrow prejudices. I had consulted with an online computer expert for months before I learned that he was fifteen years old, a mild shock that compelled me to lower the minimum age for students taking my university online computer courses.

### E-Power

*David-slays-Goliath stories abound in Netlore.*

*Small businesses triumph because of the expanded audience the Net brings them.*

*Disenfranchised people prevail over adversaries by electronically pooling their geographically dispersed resources and speaking as a collective voice.*

*Dying patients locate sources of rare blood types that keep them alive.*

*Studies of online communications track a fascinating tendency which I call "interlational"—the kind of communication that crosses social boundaries, often because the medium denies the participants the social cues that reinforce social distinctions.*

*Within a single computer conference I observed a number of examples of interlational communication: seventh graders and college professors speaking on equal terms, students sending email messages to administrators to seek redress of grievances they would be too intimidated to air in RL (real life), high school students forgoing salutations of hierarchy such as Mr. and Ms. as they interacted with their teachers on a first-name basis—something they did only in an electronic environment and nowhere else.*

In a very real sense we are distributing power to those who know what they are doing rather than reserving it for those in charge. We live in an age in which corporations are born in garages and teachers turn to ten-year-olds for help running the technology they use in their classrooms.

We see in all of this new ways to empower the underdog, a tweaking of the nose of authority which, in the land of the free and the home of brave, is hard not to admire.

Until it destabilizes us. Until it brings pornography within a few clicks of a seventh grader's computer screen when no one except the seventh grader is looking.

We live in a world of such leveraged empowerment that we are one password away from a committed hacker's shutting down the power grid or massaging the results of an election.

*We do not own our own information— everyone else owns it for us.*

We live in a world in which the power elite can turn our own medical information against us to limit our health insurance or use our financial information to retail our buying interests to catalog barons.

When those in charge use the Net to catch a murderer, we cheer. When they use it to misfile or defile our credit card information, we contemplate murder ourselves.

The technological flattening of the power structure reaches everyone, independent of cause or character, amplifying our worst and our best. As always, one person's freedom is another person's depravity, and when the little guy becomes the bad guy we tend to blame the technology rather than the person who abuses it. Our desire to have it all—the power of the Net without the abuse—is as impossible as wanting knives to be used only for peaceful purposes.

When it comes to information, it is all or nothing. After a century that saw the rise of totalitarian fascist and communist countries followed by their fall under the weight of too little publicly available information, we have made our decision: the only thing worse than living in a country that allows the creation of five hundred channels of mindless TV is living in a country that has only one.

## + EVALUATE

We ask:

- Who stands to gain most from the technology?

◪ What privacies are we losing and what abuses of freedom are we enduring to gain the power the technology offers us?

## How does the technology affect education?

### SEE

Just as cars create roads, machines create students. As STA investigators, we are concerned with the ways in which technology changes how and what we learn.

Before the advent of public education in this country, most learning was on-the-job training. From the hearth to the field to the marketplace, life was filled with direct instruction in the art of tool use.

But with the invention of the steam engine, technique became too complex to transfer on a show-me basis only. Complete transfer had to convey information about how to operate an engine as well as an understanding of how it worked so that it could be manufactured, fixed, adapted, redesigned.

That meant passing on stories about tools in the languages of math and science by using the skills of reading and writing. Schooling became the stage for these stories.

Part of schooling's appeal was that it wasn't supposed to change. As one of the first universally available institutions in United States history, mandatory public education formed a cultural bedrock that we expected to be impervious to the tremors of social evolution. Indeed, the standardization of culture that schooling achieved had a homogenizing, stabilizing effect.

*General Electric CEO Jack Welch notes that when the rate of change outside exceeds the rate of change inside, the end is in sight.*

—Quoted by Stan Davis and Jim Botkin, *The Monster under the Bed*

But even this bedrock couldn't absorb the changes brought about by the rapid evolution of information technology. Suddenly we were awash with competing information sources. Traditional schooling became a very narrow approach to the world.

With the advent of television more information became available outside the classroom than within it. And now, with the Net, information not only has exploded; it also has gone interactive. Unlike television, which fosters passive watching, the Net has begun to engage students in a self-driven process of education.

Indeed, the result of much of the technology that we create today is to put learning power in the hands of the individual. School becomes an extension of personal learning, not the other way around. As STA investigators, we are interested in how technology shifts the locus, as well as the focus, of learning.

Guidance for the new learning processes must come from somewhere. Stan Davis and Jim Botkin argue that employers can't wait for schools to provide guidance and will do so themselves. Business, they write in *The Monster under the Bed: How*

*Business Is Mastering the Opportunity of Knowledge for Profit,* "is coming to bear the major responsibility for the kind of education that is necessary for any country to remain competitive in the new economy."

A completely literate person, Davis and Botkin believe, will master not just the 3 Rs but also six other Rs needed to compete for survival: risks, results, rewards, relationships, research, and rivalry. While not new, these characteristics of the ideal employee have been brought into high resolution by the demands for a technologically capable workforce.

It is hard to argue with this perspective. Who doesn't want reward-driven, risk-taking, team-building, technologically savvy children? Still, we need to see the potential disconnective properties of this perspective, and to do so we must trudge once more up the hill to the ledge of objectivity and take a good hard look.

It may be an old-fashioned notion, but to my way of thinking school isn't supposed to prepare students for nothing but work. It is also supposed to prepare them for life. And life is — or certainly should be — more than work.

We tend to forget that most students spend only twelve to sixteen years in school and then a half-century as adults, voters, and community members. We need to concern ourselves with what we want our next-door neighbors to know and value.

Teaching people how to achieve community should be an important aim in our schools, yet the values and skills that turn a town into a community are not part of a standard business curriculum. When we fail to consider community mythology as a core subtext of the schooling process we leave our schools with no mission other than producing workers rather than citizens. Without proper management, technology becomes little more than a tool to achieve this end.

*No one ever died saying "I regret having not spent more time at the office."*
—Information Age folk saying

*Time may be money but they differ in one important respect: you can always make more money, but you can't make more time.*

## The E Word: Ethics

*Faced with the growing prospect of schools run completely on business-prep principles, a mandatory course in ethical philosophy begins to look awfully good.*

*In such a course students could discuss:*

- ◪ *How do Greek mythology and Roman ethics influence our value system?*
- ◪ *What was really bothering Hamlet?*
- ◪ *Why did the colonists feel that a free America was worth fighting for?*
- ◪ *What constitutes the common good?*

*In an age in which higher callings are being replaced by bottom lines, addressing "nonessential" curriculum items such as these begins to look less like frivolous flights of intellectual yarn-spinning and more like spiritual survival tactics.*

*What is the greatest sin committed in today's schools? Failing to help students make connections, says Dale Parnell.*

But even with a higher cause to drive us, we are left to deal with the fact that technological advancement obsolesces not only our machines but also the workers whose skills are tied to them. How do we remain work ready?

Some of the buzzwords surrounding our current educational narrative — many of which were borrowed from or co-created with the world of work — are actually quite applicable and vital: problem solving, critical thinking, lifelong learning. By now they may sound like clichés, but they describe a general set of tools that enable us to do two things: to plan the big picture and to learn as we go, drawing the picture on the fly as new information is fed to us by rapidly evolving, technologically driven learning situations.

The Industrial Age educational goal of learning and applying relatively unchanging knowledge or skill sets to fairly static situations yields to the need to be able to acquire new knowledge quickly and place it within an ever-expanding context. As STA investigators we wish to know to what extent the technology challenges us in this regard and what skills and attitudes best serve our use of the technology.

To facilitate such learning we ask the teacher to be a guide on the side rather than a sage on the stage, facilitating rather than shoveling knowledge. When the shift to this model is complete, we will value our teachers as much for their wisdom as for their knowledge. We will discover that in the Information Age we need teachers more than ever to help define the greater purposes to which learning should aspire.

## + EVALUATE

We ask:

- Do people need to be trained or retrained to use the technology?

- What role should public education play in training students to use it?

- What role should business play in training people to use it?

- To what extent does the technology facilitate self-training or promote teamwork?

- How will the technology affect our need to learn quickly, plan broadly, and adapt intelligently to evolving situations?

- What attitudes should we bring to the technology in order to learn it and use it more effectively?

### How does the technology affect other cultures?

## SEE

Much of the consumer technology that reaches developing countries comes from developed countries. *Technology transfer* is the term often used to describe the process of getting a technology from one culture to another. Given that our technologies embody purposes, goals, and political agendas, we need to ask whether the inherent qualities of a technology might throw a culture not predisposed to its use into confusion.

Paul Levinson, in *The Soft Edge: A Natural History and Future of the Information Revolution,* discusses the difference between "hard-edge" communication media, which have "an inevitable, irresistible social effect," and "soft-edge" media, which merely make events possible but "whose shape and impact are the result of factors other than the information technology at hand." Levinson uses his soft/hard distinction to consider how free will and technological determinism mix.

Television provides some of the most dramatic examples, particularly in remote areas of Alaska that are home to some of the last communities in the United States to accept it. Stories from educators who were teaching in villages when TV took hold suggest that the results were immediate and overwhelming.

Not only did the community meeting give way to primetime TV watching. Youngsters began to identify with TV characters rather than village elders as role models. The desire for subsistence food diminished, while the desire for packaged foods rose. The itch among youths to leave their villages for greener pastures often became irresistible as village life seemed more and more out of touch with their image of reality.

Such changes, whether for better or worse, are abrupt, extreme, and incontrovertible. They have a very hard edge.

The cross-cultural implications of importing technology should be of global concern because technology homogenizes a world that seems to need variety to sustain itself. Too much

*To Ellen Ullman computer systems resemble suburban developments: both take real, particular places and turn them into anyplace.*

*Before graduating from high school a young person has seen perhaps a million television commercials.*

—Ballpark figure used in communication studies

Daniel Boorstin
observes that the
supreme law of the
"republic of tech-
nology" is conver-
gence, the tendency
for everything to
become like every-
thing else.

cultural inbreeding may erase humankind's valuable differences. It may also reinforce our worst traits and weaken our ability to adapt to new situations as thinkers and problem solvers.

Technology transfer may ultimately diminish our collective, global depth and reduce the resources we can draw upon as a world system to meet the challenges that face us. The effects are quite apparent in North America, where our Native cultures are becoming so diluted that it is increasingly difficult for the larger body of humanity to appreciate and learn from their unique ways of life and their many contributions to world culture.

### Hello, World!

*Although Paul Levinson writes in* The Soft Edge *mainly about information media, his dichotomy between hard-edge and soft-edge effects is applicable to all technologies.*

*Levinson's rubrics provide a good vantage point from which to think about technology transfer in general. And nowhere does the hard edge of technology become more of a precipice than in traditional communities that adopt postmodern technologies. We could draw a legion of examples from the annals of the acculturation of indigenous peoples.*

*Technology takes us from nicety to necessity.*

*When, for instance, the Sklot Lapps of Finland traded their traditional skis and dogsleds for snow machines, as Langdon Winner points out in* Autonomous Technology, *a cascade effect was unleashed. It reduced reindeer populations, bankrupted many herders, and significantly depreciated a way of life.*

*Edmund Carpenter's* Oh What a Blow That Phantom Gave Me! *chronicles the permanent alteration of an indigenous culture in Papua New Guinea, both in habits and in consciousness, as a result of the introduction of radio, photography, and movies.*

*In my own backyard in Alaska, indigenous cultures fight the overwhelming forces of literacy, the cash economy, technology, religion, and Western learning practices. These aspects of postmodernity have not just modified their cultures but in many cases have displaced them.*

## + EVALUATE

We ask:

- What impact has the technology had on indigenous populations that have adopted it?

- How would it impact a developing country if it were introduced and used *en masse?*

- How does it homogenize culture and reduce global diversity?

### How does the technology affect technologies of the future?

## SEE

In education, business, and living, the plan's the thing. We call upon our planners to soothsay the future so that we might live proactively rather than reactively, surfboard in hand when the next wave of technologization hits.

*Planning* is often a euphemism for the enactment of a self-fulfilling prophecy. It is indeed a catch-22: without a plan, we feel we may be ambushed; with one, we ensure that the ambush will occur. A plan allows us to determine *a posteriori* whether we have failed or succeeded and gives us the scapegoats we need when we feel blindsided by technological evolution: the damned plans or the damned planners.

It's not terribly difficult to lift the veil of the future. The Chinese sages tell us that if we examine what has already transpired in a life or in a culture, we can know what lies ahead.

As with culture, so with technology: we seek to understand a technology's ancestry in order to understand its progeny. We draw a timeline from some point in the past to some point in the future and put at its center a car, a microwave, an Internet connection. Then we conceptualize the technology in terms of its overriding characteristics—the problems it solves.

Before the microwave joined the human family we cooked in firepits, fireplaces, and brick ovens and on gas and electric stoves. In this lineage we can see the evolution of the predominant characteristics of size reduction, increased speed, and control.

So what lies on the horizon for the home chef? Something smaller than a microwave? Or have we reached the limits of size reduction with the microwave as we have with the calculator? A smaller calculator would be too tiny for human digits; a smaller microwave couldn't accommodate a family-size casserole.

Will the next development be increased speed? Or is the three-minute baked potato acceptable? Will the microwave become more portable, running on batteries so we can pack it along with our laptops?

Or perhaps the microwave has fulfilled most of its evolutionary potential as a kitchen timesaver and will, like many other technologies, mutate. Will we dry our clothes by microwave? Heat our water?

We can always turn our attention to the microwave oven's subsystems or close relatives in the tecosystem. Can't we appropriate its glass or plastic to other purposes? Its timing devices?

And of course we can't help but be curious about what would happen to the microwave and all other technologies if the overriding characteristics of speed, portability, and size, along with the independence of lifestyle they facilitate, ceased to be important to us.

## + EVALUATE

We ask:

- What characteristics predominate in the technology's evolution?

- Based on the technology's ancestry, what is its most probable future?

- Does the technology make other technologies inevitable?

- Can the technology mutate or crossbreed to form new technologies?

## Answering Ourselves

It is safe to assume that my list of things to look for and evaluate in a technology is incomplete. Over the years my students have suggested other possibilities, such as the glitz factor (does it make you feel cool?), religion (does it give you something to believe in?), and the institutional imperative (does it strengthen or weaken the herd instinct?).

As STA investigators we are always vigilant, trying to see clearly not only what is before us but also whatever our myopia renders invisible. Our code of professional objectives forever remains as follows:

By bringing our background into the foreground, we see ourselves in the light of who we had forgotten we have become.

By looking at our technology, we see reflections of what we forgot we had wished for.

By listening to our technology, we can hear ourselves talking to ourselves.

By questioning where we have arrived, we can examine the purpose of our destination.

# Three

# *The Machine in the Mirror*

At the end of this process of seeing and evaluating we are required to synthesize the results of our considerations and make a judgment about the technology. As investigators in our own Science and Technology Administration, our job is to fully endorse a technology, conditionally endorse it, return it to the laboratory for modification, or reject it outright.

How do we go about making this judgment?

I require each student in my technology assessment course to assess a technology based on answers to the STA-type questions we considered in Chapter Two. The students must do the assessment in writing as concisely as possible, synthesizing and clarifying what they understand—which is, of course, the ultimate goal of most worthwhile educational activities.

Suppose you are one of my students. And suppose further that you, adopting the outlook of an STA investigator, are trying to assess a new piece of technology, perhaps with an eye to buying and using it yourself or for use in your school or business. The task of asking and answering the many possible questions in such a short space may seem impossible. How would you approach it?

Fortunately, given the great deal of overlap among the questions, it is unnecessary to address each and every one of them. In fact, I recommend considering the questions under the three broad aspects of technology assessment that I used in Chapter Two: the technology itself, the goals of the technology, and the effects of the technology.

I encourage my students to choose from a number of presentation tools: narrative, bulleted points, bubble diagrams, impact databases, T-balances. Here we will use only bulleted points and a T-balance, but I explain all of the tools in my online course, which you are welcome to view at *ivaldi.jun.alaska.edu/edtech/tat/cover/covfram.html.* The resources section of the course also offers examples of model student technology assessments. I highly recommend that you view them, too.

What technology shall we assess? Let's say that the object of your interest is an electronic book. The ebook is a splendidly mysterious and engaging technology to consider because of its wild-card status. It continues some traditions, modifies others, creates entirely new ones, and in general promises to wreak a slow havoc on life as we know it.

It is up to you to investigate the ebook's nature, goals, and impacts and ultimately to decide what kind of contribution it will make to education, business, and society in general. So, concentrating on this relatively new technology, let's revisit the three facets of technology assessment.

## The Technology Itself Revisited

For assessing electronic book technology, the best presentation tool may be a list of bulleted paragraphs. In such a list the paragraphs constitute independent observations that group well but do not necessarily follow one another logically as paragraphs in an essay do—or should.

Fluid narrative is certainly welcome, but I am more interested in not letting the form get in the way of our thinking. Bulleted paragraphs, kept as short as possible, serve this purpose well.

What can we say about "the technology itself"—the electronic book as an artifact that is both tool and machine, with characteristics of involvement, extension, limitation, capacity, and dependability?

- *Features.* The concept of a virtual, "intelligent" book is not new. Many teachers use online, downloadable textbooks, or hyperbooks. The limitations inherent in hyperbooks appear in hindsight to have made ebooks inevitable. Because hyperbooks download directly to the reader's hard drive, portability can be a problem. The reader who wishes to use the hyperbook in more than one location must carry around a portable computer. The ebook is much lighter and more durable than a laptop, and it focuses on doing just one thing: being a book. It approximates a conventional paper-based book, substituting the screen for the page. You, the reader, can turn pages, refreshing the screen with new data, but you can't spread out several full pages side by side or physically tab pages for purposes of cross-referencing. That is not a problem when you read a bound book, whose pages can be easily bookmarked. But it can be an impediment if you wish to read project reports or technical data. To compensate, the ebook offers not only bookmarking but also the value-added processing capabilities of searching and cross-referencing within and outside of the text. It also allows you to personalize data by making notes, amending the text with pictures and other media, and changing fonts and font sizes to improve readability. An ebook can store many books simultaneously. That lets you do quick cross-referencing while reducing the overall bulk and weight of the conventional books you would otherwise need to handle. You will be able to take your personal library with you wherever you go.

- *Hardware.* The electronic book is made of plastics and microcircuitry, with a range of finishes expected to be available, from punk to professional. Whereas conventional book technology works under almost any conditions so long as the reader has light to see by, an electronic book depends on a power source and technical support. The average reader can't fix it as one might repair a torn page, so the technology requires outside maintenance. Like such gadgets as the Walkman and the portable CD player, this technology encourages repurchasing. Depending on the ebook's price and the heft of your pocketbook, you may choose to throw it away and buy a new one rather than trying to have it fixed.

- *Software.* Paper-based books are complete entities; they are software wedded to hardware, to paraphrase Paul Levinson in *The Soft Edge.* But because of that they are also static. Electronic books are more dynamic, like a Nintendo machine. You don't need to keep reading the same old text. Much as you can change the Nintendo game you're playing, you can buy and download new texts into your ebook, as well as purge old ones without contributing them to a secondhand bookstore.

- *Tool and machine.* An electronic book is an interesting combination of tool and machine. It provides the mechanical qualities we have come to expect from an author, the techniques that lead you, the reader, through a sequential story. It also provides very tool-like qualities that allow you to direct the reading process by using the value-added processing capabilities already mentioned: bookmarking, searching, and cross-referencing.

- *Reader extension.* The electronic book extends us in many of the ways a paper-based book extends us. By preserving forever the voice and views of an author we probably won't meet, an author who lives far away or died long ago, both kinds of book extend our eyes and our ears. But the ebook adds a fluidity and adaptability that is quite new. It allows us to interact with contemporary authors via email—and maybe even have an influence on the next books they write. And electronic fiction can be programmed with non-presumptive branching so that the reader can easily determine the turns of the plot as well as a book's ending. That can solidify and expand the partnership between author and reader, blurring the lines between the two.

- *Physical stress and training.* Like any computer screen, the electronic book forces the reader to forgo a static, high-resolution page for one that is harder to see and demands much more eye-muscle work—and more eyestrain. After four decades of wretched screen technology, we are still waiting for the research and development people to rescue our eyes from this hazard. And what untold havoc will gripping a hard ebook wreak on our hands and bodies that gripping a paperback doesn't? As for training, while the operation is fairly self-explanatory, using ebooks properly requires at least some training, and you must add this need into the equation.

Before ending your revisit to "the technology itself," you might want to consider a conundrum that applies to the electronic book as it does to computerized reading in general. Being a conundrum, it has no easy answer.

Were we meant to read *Walden* in a linked environment, darting from Henry Thoreau's text to Ralph Waldo Emerson's reactions to it, then to a biography detailing Thoreau's living expenses during the period he was writing *Walden,* then on to who knows where else before returning to the original text — if, indeed, we ever get back to it? Does this procedure do harm not only to Thoreau's intent but also to our understanding of his message? Or will such interactions increase our understanding and enrich our *Walden* experience?

## The Goals of the Technology Revisited

If, as Abraham Maslow said, the world looks like a nail to someone with a hammer, what does it look like to someone with an electronic book? What behaviors are implicit in this new medium?

The medium of the printed book reshaped so much of humankind's experience of the world that McLuhan and many other communication theorists dedicated their lives to explaining the phenomenon of literacy and its implications. To them, embedded in books, irrespective of content, was a cultural shift in which individual, silent reading replaced group storytelling, forcing literate cultures into a linear, visual world and away from a holistic one in which the omnipresent senses of touch, smell, and hearing dominated. The result was our detribalization, the fragmentation of society into individuals, and the coalescence of cultures around linguistic bases, which we call nations.

Surely many of the qualities of the book will rub off on its electronic offspring. But what is different about electronic books? What additional inherent biases can we see?

- *Associative reading.* Electronic books will be built so as to optimize navigation features. And because the ebook lets us link and cross-reference, we will do so. We will feel compelled to explore and grow into the medium's capabilities. That means we will begin to read and think associatively rather than sequentially—I link, therefore I am. The very nature of book reading will change. Readers no longer will be led by an author on a predetermined route but will participate actively, even guiding the process.

- *Theft.* We will feel possessive about ebooks. They will require protection from theft until they become as inexpensive as paper-based books. When was the last time you locked your car because you were afraid someone would steal your book?

- *Book buying.* Our book-buying practices will change dramatically. Many ebook texts may even be free for the downloading—as long as we are willing to accept the advertisements that come with them. These ads will be pointcasted, their content based on the nature of the text, the personal data we feed to the online book provider, or both. We may lose our ability to separate commercial from artistic concerns while gaining almost unlimited access to texts on the Web. But libraries may make commercial-free books available online for short periods, after which they become unreadable; that would render the overdue library book obsolete. For publishers the challenge will be to package the text with Web links, associated books and resources, author biographies, even directions for buying related items.

- *Time shift.* Using electronic books will increase the amount of time we spend online and reduce the time we spend browsing in bookstores and sipping espresso at adjoining cafés.

- *Legal issues.* Ebooks will bring new ethical and legal concerns. No one considers it illegal to lend paperback books to friends who can't afford to buy their own. But what if the books are electronic? New discussions will emerge about who owns information and who must pay for it.

- *Flexibility.* Public and private bookshelves will continue to house traditional books that are aesthetically pleasing and whose content is enhanced by print: atlases, art books, and others requiring high resolution. Leisure books, textbooks, and manuals will go electronic. Schools, businesses, and others that routinely buy books with short half-lives will start using ebooks as soon as it becomes feasible to do so. In this way they will not only save money and warehouse space but also gain flexibility in the adoption of reading materials. Ebooks will finally make textbooks available for just-in-time learning.

- *Multiple uses.* How we interpret our electronic books will depend on their features. The technology could replace not only our printed books but also our daily planners, phone directories, and paper-based notebooks. I began carrying a notebook eighteen years ago to record the flow of activities in my professional life and now have a box full of notebooks. The electronic book, if it doubled as a writing pad, could reduce this weighty collection to a single storage disk. With the addition of minimal computing abilities, such as Web searching and word processing, the ebook could become all the computer most people need, making the work world far less desk-based and much more mobile. With sound, video, large fonts, and other features it could be a useful technology for physically challenged persons.

## The Effects of the Technology Revisited

Every technology both connects and disconnects, bringing us both gains and losses. Using a T-balance is a good way to demonstrate this truism.

A simple T-balance has countering statements on each side of a T-shaped divider. To examine the effects of electronic book technology, we will use a value-added T-balance of sorts, with a trait column to the left of the T-balance.

In the trait column we list the impact areas we discussed in "The Effects of the Technology" section in Chapter Two. In the T-balance we list ways in which ebook technology makes connections (left-hand column) or disconnections (right-hand column) in these impact areas.

There is a great deal of overlap among the areas, and the assignment of aspects of the technology to one trait versus another is often arbitrary. But the structure of the T-balance will help in achieving your goal as an investigator for the STA: sorting out and synthesizing the multitude of issues that surround the ebook and communicating the results effectively to others.

|  | CONNECTION | DISCONNECTION |
|---|---|---|
| **TRAIT** |  |  |
| *The environment* | Ebooks save lots of paper and many trees; reduce the physical space that now holds printed books, making it available for other uses; greatly diminish the resource-intensive infrastructure supporting printed books. | Ebooks create toxic waste byproducts in producing semiconductors; encourage obsolescence that produces a plethora of non-biodegradable electronic trash; greatly increase the resource-intensive infrastructure needed to support them. |
| *The human body* | Ebooks amplify and extend our eyes, ears, and nervous system; reduce the burden of carrying heavy printed books. | Ebooks increase eyestrain due to poor resolution; may induce muscle problems due to gripping. |
| *Social relationships* | Ebooks on networks make reading a collaborative experience; popularize chat groups about books. | Ebooks replace a relatively cheap commodity with an expensive one, limiting access to those who can afford them; encourage theft. |
| *Work* | Ebooks create new jobs for writers, artists, and others who create and produce texts and graphics; provide new opportunities in ebook technology and in self-publishing. | Ebooks displace workers in print book production; threaten workers and management in traditional publishing. |
| *Self-relationships* | Ebooks enhance self-sufficiency through self-paced, self-guided learning with new navigation and annotation tools; encourage self-exploration of ideas and concepts; personalize the reading experience. | Ebooks sacrifice the depth of printed books for the breadth of electronic options; may prove to be just another electronic distraction — Nintendo for the reading addict. |
| *Power structure* | Ebooks, made of bits rather than atoms and thus more fluid than paper-based books, accelerate book-swapping; provide more reader control, with a diminished possibility of | Ebooks make data so fluid that sharing becomes easy, copyright becomes elusive, and compensating authors becomes difficult; greatly reduce the role of traditional |

| CONNECTION | DISCONNECTION |
|---|---|
| censorship; reduce the broadcast feature of printed books by enhancing reader interaction. Publishers may respond with technology that makes sharing difficult; some may offer ebook club memberships as a friendlier alternative. | publishers; limit the "right to e-read" to those who can afford it. |

|  | CONNECTION | DISCONNECTION |
|---|---|---|
| *Education* | Ebooks require readers to be trained to operate the hardware and optimize the features, to think more associatively and less sequentially, and to read more cooperatively and less individually. | Ebooks may force institutions to play catch-up if they don't foresee the technology's impacts; may create social dissonance as a result of incompatibilities in information processing practices within and outside educational environments; may replace equal distribution of textbooks with privileged use of the new technology based on family resources. |
| *Other cultures* | Ebooks are potentially more adaptable than printed books to different modes of thought and different reading styles; could have built-in translators and culture-specific annotations. | Ebooks have an initial cost that will prohibit readers in developing economies from using them, thereby reinforcing the have/have-not social structure; may homogenize views of producing cultures and consuming cultures. |
| *Future technologies* | Ebooks of the future could offer greater support by online services, screens with better resolution, book bodies made of more flexible material that bends the way paperbacks do, special finishes to approximate the tactile sensation of holding a book, multimedia features. | Ebooks promise future enhancements that ultimately will make sequential reading obsolete. |

## The Judgment

As investigators for the Science and Technology Administration, having duly weighed the electronic book technology itself along with its goals and its effects, we must now render our judgment. Is the ebook worthy of our approval for purchase and use?

Yes, we grant the electronic book our approval. It offers opportunities to share and personalize reading in ways that hold a great deal of promise for education, self-directed learning, and the physically challenged. And it promises to save a precious resource: trees.

Our approval, however, is contingent. It depends on several factors.

First, the ebook must be made as biodegradable and recyclable as possible.

Second, it needs to be designed so as not to foster eye fatigue or muscle problems.

Third, ebook manufacturers must support efforts to provide their products to public libraries and developing countries so that persons who can't afford them can use them.

Fourth, schools and businesses need to support training efforts to optimize people's use of ebooks while preserving the depth and strengths of printed books.

# Last Words

A circular diagram similar to the one I ask my students to draw in order to see the big picture (described in "The Effects of the Technology" section in Chapter Two) has gained currency in my own thinking.

In a single illustration this diagram captures the way we have traditionally approached the use of technology in our schools, businesses, and communities. It also illustrates its inverse — how we *should* approach it.

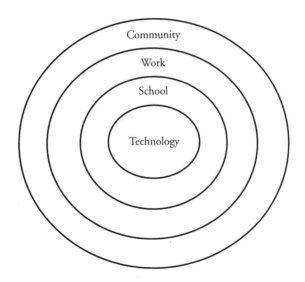

## Seeing Is a Matter of Vision

We return to the basic issue that drives this book: just what is it that we want when it comes to technology? More specifically, what is it that we want from our schools in regard to technology? What attitudes about technology do we want to instill in our children? What kind of adults do we want our children to be when they emerge from their twelve to sixteen years of education and enter the full life of the workplace and community?

Suppose we could practice zero-based education planning, which would allow us to rebuild an educational system from the ground up without any strings to the past. What would it look like? And what would we have our children become?

The diagram, with its four circles labeled "Technology," "School," "Work," and "Community," captures quite well the story of the first twenty glorious and bumbling years of the introduction and use of new electronic technologies in education, particularly in the United States.

*The drive to saturate classrooms with computers is marked by faith and fear, says William L. Rukeyser: "Faith that computer skills guarantee economic success.... Fear that we'll miss the boat if we don't rush our decisions."*

*—Quoted by Stan Davis and Jim Botkin, The Monster under the Bed*

*Circle 1: Technology.* In the United States we worked from the inside of this diagram outward, beginning with the technology and working our way outward to the community. We had no choice—or so it seemed.

In the early 1980s computer technology was suddenly upon us and quickly proliferating. Apple IIe's sprouted in classrooms all over the country. We had to accommodate the presence of microcomputers before we could see their implications. We knew that an information technology revolution was at hand but had no idea where it would take us.

Our greatest fear was of being left behind as the future engulfed us. Most educational technology planners felt we were already late for the party and might not get there before the food was gone.

Compelled to outfit our schools, we started at the center of the circle and moved outward. We bought computers and waited for a miracle to turn them into something other than programming machines that were useless to the average teacher in the classroom.

*Circle 2: School.* When users switched on their Apple IIe's, which dominated educational computing through most of the 1980s, they were greeted by the BASIC programming language—not a pleasant way to start the day. Early IBM computers weren't much better, landing users at a mysterious prompt and requiring them to navigate in an environment only slightly less alien than programming.

*How different the computer revolution would have been had the first personal computers been word processors rather than programming machines!*

*—Information Age folklore*

With the arrival of AppleWorks software, running on the Apple IIe, the world of educational technology gained its first real productivity tool for teachers and students. AppleWorks combined database management, spreadsheets, and word processing in an environment that was easy to use and adaptable to a wide variety of content areas. This suite of programs is often referred to collectively as tool software.

Then and only then did we move fully into the second circle in the diagram, applying computers to the art of teaching and learning.

*Circle 3: Work.* The new technology moved from school to work when people recognized that computing skills were no longer just for information technology specialists but were vital in every field of endeavor.

*If culture is the way we organize and understand our actions, Michael Goldhaber contends, then in many work settings technology very nearly is culture.*

Almost overnight a command of information technology became an essential skill for nearly everyone entering the world of work. A student who could compute could work, the thinking went, and a working community was a stable, productive, vibrant community. Bond issues, tax increases, budget realignments, and other forms of support for technology in the schools began to gain public favor.

*Circle 4: Community.* And now, finally, we're asking how we can use our new technologies to make our communities more livable. But we have only begun to ask; we haven't yet moved fully into the fourth circle.

Imagine what would happen if we reversed this process, working from the outside of the circle inward. Suppose our educational technology planning process proceeded by addressing four essential questions:

*The value of a new technology, says Nicolas Jéquier, lies not only in its economic viability and its technical soundness but also in its adaptation to the local social and cultural environment.*

- ◪ First, what do we want our community to be like?
- ◪ Second, what kind of work do we need to provide to create the community we envision?
- ◪ Third, what kind of education will this kind of work require?
- ◪ And finally, how can technology help us reach our educational goals?

That order of thinking seems obvious. Yet we were so distracted by present problems and so overwhelmed by fear of the future that we failed to notice it.

## Becoming Involved

For most people the issue ultimately becomes one of involvement. If we want our children to cultivate stage 4 literacy about technology, how do we go about it? What can we as teachers, parents, businessmen, and citizens do to promote a balanced and informed view of the impact of technology in our schools?

The current debate over educational standards and assessment provides a good point of entry into the issue of using technology wisely in our schools—and ultimately in our communities.

Many states have developed educational standards for major content areas, including educational technology. Standards typically consist of lists of competencies in each curricular area which a state's department of education, through its standards development team, considers essential to a child's education.

Standards ostensibly identify what society in general considers essential knowledge for an educated person. But in fact much of the drive for standards has come from employers who want our schools to produce graduates they can hire—potential workers with basic literacy skills.

Many educators feel that standards are a bittersweet addition to their teaching considerations. They see standards as a source of both consolation and consternation.

On one hand, the standards provide a socially sanctioned perspective of what is important for children to learn and accomplish in school. They help educators gauge and modify course contents and teaching practices to reflect public concerns. They help guide public education in directions that a majority of citizens can support.

On the other hand, standards can constrain teachers and students by not allowing for innovative curricula and pedagogy, unique situations and learning styles, or students' special needs. They can make unrealistic demands of geographically or financially disadvantaged districts. They can be culturally biased.

While elements of truth exist in both perspectives, standards are undoubtedly beneficial in one regard. They provide a point for citizens to enter into a public dialogue about improving education, a dialogue that has local and national implications.

In my home state of Alaska the process of developing education standards took two years and involved many people from many walks of life: K–12 educators, business people, parents, citizen advocates, university educators, government officials. You can find the results of these efforts on the Alaska Students Content Standards home page *(www.educ.state.ak.us/ ContentStandards/home.html)*, which is part of the Web site maintained by the Alaska Department of Education.

Within this online document are Alaska's content standards for technology *(www.educ.state.ak.us/ContentStandards/ Technology.html)*. The last technology standard listed there, standard E, relates closely to the subject of this book. It reads:

> A student should be able to use technology responsibly and understand its impact on individuals and society.
>
> A student who meets the content standard should be able to:
>
> ◪ Evaluate the potentials and limitations of existing technologies.
>
> ◪ Discriminate between responsible and irresponsible uses of technology.
>
> ◪ Respect others' rights of privacy in electronic environments.
>
> ◪ Demonstrate ethical and legal behavior regarding intellectual property, which is the manifestation of an original idea, such as computer software, music, or literature.
>
> ◪ Examine the role of technology in the workplace and explore careers that require the use of technology.

*Goals not written down are just wishes.*

—Anonymous folk wisdom

- Evaluate ways that technology impacts culture and the environment.
- Integrate the use of technology into daily living.
- Recognize the implications of emerging technologies.

Extensive Web searching completed by my academic department in August 1998 shows that twenty-one states had standards specifically addressing the area of ethical and responsible use of technology and twenty-two had standards that were integrated into other content areas. But eighteen states had no standards in this area at all. (The total is more than fifty because there is overlap; some states address standards both separately and in content areas.)

Standards are always being revised and expanded. If you live in a state that has inadequate standards relating to technology or no standards at all, you don't need to accept the status quo. You can work for change, even if it requires petitioning the state school board. And you should feel free to do so; it is funded with your money.

Another point of entry into the public dialogue about technology is the debate over putting standards into practice and how to assess their implementation. How do we move from standards theory to standards reality?

There is no time like the present to help shape this debate. It is just starting to roll and promises to be spirited, complex, and ultimately transformative, no matter what happens. It is carried on largely by public groups funded with public money. Giving your input is a democratic privilege and responsibility.

The final result, one hopes, will be a redefinition of the successful student as one who masters curriculum content, achieves technological dexterity, and becomes socially aware. And, let us also hope, the successful student will be one who

*Our deepest fear is not that we are inadequate; our deepest fear is that we are powerful beyond measure.*

—Nelson Mandela

knows not only how to use technology but also when to use it and when not to use it. The successful student, finally, will be one who understands that technology can both disconnect us from and connect us to opportunities.

Our new technologies have placed us at the beginning of a new era of learning. If we're driven by the right spirit to apply these technologies wisely, we can turn our schools, businesses, and communities into anything we want them to be. We can redefine what it means to be educated in bold new ways. We can see technology anew.

# Bibliography

Ivan Amato. *Stuff—the Materials the World Is Made Of.* New York: Basic Books, 1997.

Sven Birkerts. "The Electronic Hive: Refuse It." In Bill Henderson, editor. *Minutes of the Lead Pencil Club: Pulling the Plug on the Electronic Revolution.* Wainscott, N.Y.: Pushcart Press, 1996.

Daniel J. Boorstin. *The Republic of Technology: Reflections on Our Future Community.* New York: Harper & Row, 1978.

Albert Borgmann. *Technology and the Character of Contemporary Life.* Chicago: University of Chicago Press, 1984.

Ernst Braun. *Wayward Technology.* Westport, Conn.: Greenwood Press, 1984.

John Brooks. *Telephone: The First Hundred Years.* New York: Harper & Row, 1976.

James Burke. *Connections.* Boston: Little, Brown, 1978.

Edmund Carpenter. *Oh, What a Blow That Phantom Gave Me!* New York: Holt, Rinehart and Winston, 1973.

James Cary. "Communication and the Democratic Process." In *Critical Connections: Communication for the Future.* Office of Technology Assessment, U.S. Congress. OTA-CIT-407. Washington, D.C.: Government Printing Office, 1990.

Arthur C. Clarke. *How the World Was One.* New York: Bantam Books, 1992.

Ruth Schwartz Cowan. *More Work for Mother: The Ironies of Household Technology from the Open Hearth to the Microwave.* New York: Basic Books, 1985.

John Davies. *Educating Students in a Media-Saturated Culture.* Lancaster, Pa.: Technomic, 1996.

Stan Davis and Jim Botkin. *The Monster under the Bed: How Business Is Mastering the Opportunity of Knowledge for Profit.* New York: Simon & Schuster, 1995.

Michael L. Dertouzos. *What Will Be: How the New World of Information Will Change Our Lives.* San Francisco: Harper Collins, 1997.

Jacques Ellul. *The Technological Society.* New York: Knopf, 1973.

Stephen Fox. *The Mirror Makers: A History of American Advertising and Its Creators.* New York: Vintage Books, 1985.

Robert Frost. "The Constant Symbol." *Atlantic Monthly,* October 1946.

Howard E. Gardner. *Frames of Mind: The Theory of Multiple Intelligences.* New York: Basic Books, 1993.

Michael Goldhaber. *Reinventing Technology: Policies for Democratic Values.* New York: Routledge & Kegan Paul, 1986.

Bill Henderson, editor. *Minutes of the Lead Pencil Club: Pulling the Plug on the Electronic Revolution.* Wainscott, N.Y.: Pushcart Press, 1996.

Nicolas Jéquier, editor. *Appropriate Technology: Problems and Promises.* Paris and Washington: Development Center of the Organization for Economic Cooperation and Development, 1976.

Susan Jonas and Marilyn Nissenson. *Going, Going, Gone: Vanishing Americana.* San Francisco: Chronicle Books, 1998.

Friedrich George Junger. *The Failure of Technology.* Chicago: Regnery, 1956.

R.D. Laing and A. Esterson. *Sanity, Madness, and the Family: Families of Schizophrenics.* Middlesex: Penguin Books, 1974.

Joseph Lanza. *Elevator Music: A Surreal History of Muzak, Easy-Listening, and Other Moodsong.* New York: St. Martin's Press, 1994.

Paul Levinson. *The Soft Edge: A Natural History and Future of the Information Revolution*. New York: Routledge, 1998.

Jerry Mander. *Four Arguments for the Elimination of Television*. New York: Quill, 1978.

Jerry Mander. *In the Absence of the Sacred: The Failure of Technology and the Survival of the Indian Nations*. San Francisco: Sierra Club Books, 1992.

Marshall McLuhan. *Culture Is Our Business*. New York: Ballantine Books, 1972.

Marshall McLuhan. *Understanding Media: The Extensions of Man*. New York: New American Library, 1964.

Marshall McLuhan and Bruce R. Powers. *The Global Village: Transformations in World Life and Media in the 21st Century*. New York: Oxford University Press, 1989.

Joshua Meyrowitz. *No Sense of Place: The Impact of Electronic Media on Social Behavior*. New York: Oxford University Press, 1986.

Elting E. Morison. *Know-How to Nowhere: The Development of American Technology*. New York: Basic Books, 1975.

Lewis Mumford. *Technics and Civilization*. San Diego: Harcourt Brace Jovanovich, 1963.

Nicholas Negroponte. *Being Digital*. New York: Knopf, 1995.

Donald Norman. *Things That Make Us Smart*. New York: Addison-Wesley, 1993.

Vance Packard. *The Pyramid Climbers*. New York: McGraw-Hill, 1962.

Vance Packard. *The Status Seekers: An Exploration of Class Behavior in America and the Hidden Barriers That Affect You, Your Community, Your Future*. New York: David McKay, 1959.

Dale Parnell. *Why Do I Have to Learn This?* Waco, Tex.: Cord Communications, 1995.

Henry Petroski. *The Evolution of Useful Things: How Everyday Artifacts—from Forks and Pins to Paper Clips and Zippers—Came to Be as They Are*. New York: Vintage Books, 1993.

Neil Postman. *The End of Education: Redefining the Value of School*. New York: Knopf, 1995.

Neil Postman. *Technopoly: The Surrender of Culture to Technology.* New York: Vintage Books, 1993.

Stephen Rosen. *Future Facts: A Forecast of the World as We Will Know It before the End of the Century.* New York: Simon & Schuster, 1976.

Jeffrey Rothfeder. *Privacy for Sale: How Computerization Has Made Everyone's Private Life an Open Secret.* New York: Simon & Schuster, 1992.

Kirkpatrick Sale. *Rebels against the Future: The Luddites and Their War on the Industrial Revolution — Lessons for the Computer Age.* Reading, Mass.: Addison-Wesley, 1996.

Secretary's Commission on Achieving Necessary Skills. *What Work Requires of Schools: A SCANS Report for America 2000.* Washington, D.C.: U.S. Department of Labor, June 1991.

Don Tapscott. *Digital Economy: Promise and Peril in the Age of Networked Intelligence.* New York: McGraw-Hill, 1996.

Don Tapscott. *Growing Up Digital: The Rise of the Net Generation.* New York: McGraw-Hill, 1998.

Edward Tenner. *Why Things Bite Back: Technology and the Revenge of Unintended Consequences.* New York: Knopf, 1996.

Sherry Turkle. *Life on the Screen: Identity in the Age of the Internet.* New York: Simon & Schuster, 1995.

Ellen Ullman. *Close to the Machine: Technophilia and Its Discontents.* San Francisco: City Lights Books, 1997.

Kurt Vonnegut. *Wampeters, Foma, and Granfalloons.* New York: Dell, 1992.

Robert H. Waterman, Jr. *Adhocracy.* New York: Norton, 1993.

Joseph Weizenbaum. *Computer Power and Human Reason: From Judgment to Calculation.* New York: W.H. Freeman, 1976.

Margaret Wheatley. *Leadership and the New Science.* San Francisco: Berrett-Kohler, 1994.

Leslie A. White. *The Science of Culture: A Study of Man and Civilization.* New York: Grove Press, 1949.

Richard Wilhelm, translator. *I Ching or Book of Changes.* Rendered

into English by Cary F. Baynes. Third edition. Princeton,
N.J.: Princeton University Press, 1967.

Marie Winn. *The Plug-In Drug: Television, Children, and the Family.*
Revised edition. New York: Viking Penguin, 1985.

Langdon Winner. *Autonomous Technology: Technics-Out-of-Control as
a Theme in Political Thought.* Cambridge, Mass.: MIT Press,
1977.

Langdon Winner. *The Whale and the Reactor: A Search for Limits in
an Age of High Technology.* Chicago: University of Chicago
Press, 1986.

Tom Wolfe. *The Pump House Gang.* New York: Bantam, 1972.